Emerging Energy Frontiers in Research

Topical Conference at the 2009 AIChE Spring National Meeting

Tampa, Florida
26 – 30 April 2009

ISBN: 978-1-61567-257-8

Printed from e-media with permission by:

Curran Associates, Inc.
57 Morehouse Lane
Red Hook, NY 12571

CURRAN ASSOCIATES INC. proceedings.com

Some format issues inherent in the e-media version may also appear in this print version.

Printed by Curran Associates, Inc. (2009)

For permission requests, please contact AIChE
at the address below.

AIChE
3 Park Avenue
New York, NY 10016-5991

Phone: (203) 702-7660
Fax: (203) 775-5177

www.aiche.org

Additional copies of this publication are available from:

Curran Associates, Inc.
57 Morehouse Lane
Red Hook, NY 12571 USA
Phone: 845-758-0400
Fax: 845-758-2634
Email: curran@proceedings.com
Web: www.proceedings.com

TABLE OF CONTENTS

5a USDA ARS Bioenergy Research Strategy

Robert L. Fireovid, USDA Agricultural Research Service, Room 4-2166, 5601 Sunnyside Ave, GWCC-BLTSVL, Beltsville, MD 20705

The growth and long-term viability of bioenergy production in the Nation are impeded by a number of technical and commercial barriers. ARS addresses only technical barriers and does so by leveraging its strengths and unique capabilities to (1) pursue technical

barriers that can be overcome by ARS resources and (2) provide substantial and unique contributions towards the Nation's bioenergy goals.

Web Page: www.ars.usda.gov/research/programs.htm

5b DOE "Advanced Biofuels" Strategy

Valerie Sarisky-Reed, Office of the Biomass Program, US Department of Energy, 1000 Independence Ave, SW, EE-2E, Washington, DC, DC 20585

The Energy Independence and Securities Act of 2007(EISA) establishes a very aggressive renewable fuels standard (RFS) that calls for 21 billion gallons of advanced biofuels for transportation use by 2022. While cellulosic ethanol is likely to play a significant role in meeting the EISA goals, a number of other advanced biofuels show promise as being able to contribute. The suite of biofuels includes gasoline and diesel replacement fuels that are essential hydrocarbons made from biomass rather than petroleum and include algae as a potential feedstock for production. These fuels are in various stages of development by both the public and private sector. The purpose of this presentation is to discuss the advantages and disadvantages of these various fuels and the DOE strategy to pursue cost effective technologies to produce them commercially.

Web Page: www.eere.doe.gov

5c Emerging Technologies and Alternative Energy: a Perspective of a Petroleum Exporting Jurisdiction

Dzung M. Nguyen, Eddy Isaacs, and Surindar Singh. Emerging Energy Technologies, Alberta Energy Research Institute, 500 Phipps McKinnon Building, 10020 - 101A Avenue, Edmonton, AB T5J 3G2, Canada

Alberta has the world's 2nd largest crude oil reserves and is the largest foreign supplier of crude oil to the United States* . Alberta's current prosperity in energy is largely based on sustained aggressive public and private investments in research and development and field trials, which led to many new oil/gas recovery processes and enabling technologies. This is especially true in oil sands where Alberta has over 70% of the world's resources. These experiences have shown that governments have a key role to play in the development of new technologies. They can make innovation happen, across the full energy technology spectrum, from applied basic research and technology development to engineering design, pilot and pre-commercial demonstration projects.

The integration of the North American economy presents both opportunities and challenges for Alberta. The government is committed to ensuring that its energy sector remains viable and attractive for investment. Strategically positioning itself in a future where there is a transition from global dependence on fossil fuels to Alternative Energy presents a major challenge among many.

The Alberta Energy Research Institute (AERI) serves as the energy technology arm of the Government of Alberta. Its mission is to accelerate the development of clean energy technologies through research, technology and innovation. AERI has a mandate to position Alberta for the future in energy; add value to its energy resources; encourage and invest in technology and innovation; and coordinate energy and environmental R&D across government agencies.

AERI's Emerging Energy Technologies strategic intent is to foster and support the adoption, adaptation and/or development of emerging technologies for immediate application to the problems and challenges of the current fossil world, then leverage the resulting innovation and knowledge base to building a stronger economic foundation for that of an Alternative Energy future. Examples are Underground Coal Gasification for hydrogen production, nano-structured materials for gas separation/storage and geothermal recovery from abandoned oil/gas wells.

AERI has extensive experience in investing both in applied research, field pilots and in demonstrations of Next-Generation energy technologies. It includes:

• Roadmapping: The importance of identifying gaps in both technology and policy.

• Focus: The need to focus on high impact-technology platforms.

• Balance: The need to maintain a balanced portfolio of investments.

• Collaboration: The importance of joint government-industry investments to share the risk of pilot and demonstration projects.

• Transparency: The need for diligent screening and selection of technologies and partners, especially in areas of technology implementation – having many partners is not necessarily desirable.

• Sustained Support: Progress can only be made through "learning by doing", and persistence and long-term sustained effort are the keys to success.

Examples of AERI current investment in emerging energy technologies, such as advanced gasification, nano-structure materials, etc. and their implications from an Alternative Energy perspective will be presented. A summary of available R&D funding in energy and environment, along with the $2 billion investment for research and implementation of carbon capture and sequestration will also be highlighted.

* US Energy Information Administration, December 2008

From Forests to Clean Fuels – Biofuels Research at the USDA Forest Service

The USDA Forest Service is charged with caring for one our Nations most important strategic assets, our forests and grasslands. Our mission is to *sustain the health, diversity, and productivity of the Nation's forests and grasslands to meet the needs of present and future generations.* The Forest Service is the only federal agency with mission goals that encompasses research from feedstock to conversions technology. Our research strategically targets underutilized residual biomass remove from forest fire protection operations, forest restoration operations and forest health improvement operations. Key programs in our biofuels research include feedstock production, feedstock harvesting, collection and transportation, conversion technologies and economic assessment. The results of our research in biofuels contribute to our nation's alternative fuel goals set by our President and Congress, and provide incentives for forest managers to remove unwanted biomass in forests that will positively impact the health and sustainability of our forests.

5e Atlantica Region's BioEnergy Perspective

Hemant P. Pendse, Managing Director, Forest Bioproducts Rsearch Initiative (FBRI), University of Maine, 5737 Jenness Hall, Room 115, Orono, ME 04469-5737, Neil Bernard, Ceres BioVentures, Ltd., Surey, United Kingdom, and Thor Oleson, Atlantica BioEnergy Task Force.

The Atlantica BioEnergy Task Force addressed a unifying challenge – how to ensure the future sustainability and prosperity of the forest products industry across three jurisdictions, New Brunswick, Nova Scotia and the State of Maine, which comprise the Atlantica Region (the "Region"). The Task Force is comprised of regional government, industry, utilities, federal and regional organizations and post secondary institutions. A comprehensive best practices study (the "Study") was identified as a foundational next step for the Task Force. As a result, PricewaterhouseCoopers LLP was commissioned to complete the Study as a cornerstone to the Task Force's efforts that are focused on understanding the challenges and opportunities facing the Region. The Task Force identified opportunities that will promote the use of renewable energy technologies using the Region's forest products industry infrastructure. There are several opportunities to revitalize the forest industry and its competitiveness by making use of the available wood feedstock—a renewable asset—for the production of bioenergy, biofuels and biochemicals and thus enhance sustainable economic development within the Region.

Governments in the three jurisdictions in the Region are currently faced with the growing need to control the source, cost, and environmental effects of energy. To decrease the reliance on carbon sources, reduce costs and to increase energy independence, many jurisdictions are developing and implementing comprehensive public policy frameworks to manage and incentivize changes in the energy supply. The public policy frameworks that are being developed are aiming to achieve several objectives including maximizing the economic development benefits from the forest, encouraging the development of reliable and diversified supplies of renewable energy and contributing to climate change initiatives such as reducing greenhouse gas emissions. Four technological opportunities were evaluated considering sustainable use of resources, current regional operations, maturity of new technologies, the current regulatory environment and the economic impact.

Gas Diffusion Layer Fiber Size Distribution and Diffusion Regime Effects on PEM Fuel Cell Performance

Manolis M. Tomadakis, Florida Institute of Technology, Melbourne, FL 32901
Katie Pentas Fuselier, Intersil Corp., Melbourne, FL 32902

The effective diffusivity of fuel cell gas diffusion layers (GDL) of bimodal (two-diameter) fiber size distribution was computed through random walks in the bulk, transition and Knudsen diffusion regime, for both in-plane and transverse flow. The bulk regime effective diffusivity was used to derive also the formation factor (hence thermal and electrical conductivity) and viscous permeability of the gas diffusion media. All transport properties were applied into a polymer electrolyte membrane (PEM) fuel cell model to better describe transport of the feed gases through bimodal fibrous gas diffusion layers. The model was used to evaluate the effects of key GDL and process parameters on fuel cell performance. The effective diffusivities computed for the bimodal GDL media in all diffusion regimes practically coincide with those for unimodal (one-diameter) media, for porosities above 0.5. This is not the case at lower porosities, where bimodal GDL structures exhibit higher diffusivities than the unimodal ones, due to a significantly lower percolation threshold (0.045 vs. 0.11). This difference in the percolation porosity gives rise to a significant difference in the viscous permeability as well, with the bimodal fiber structures found to be much more permeable to viscous flow than the unimodal ones in the low porosity range. The PEM fuel cell model predictions were evaluated at three GDL porosity levels, and the best results were achieved with the intermediate porosity of 0.4. The high GDL porosity tested, 0.7, led to a slightly lower prediction for the power density, but the low porosity of 0.12 resulted in a dramatic reduction in the value of that key fuel cell performance parameter. Bulk diffusion was found to be the dominant mass transfer mechanism in the fuel cell GDL. Transition or Knudsen diffusion can be established when using an extremely low pressure or fiber size, both having a detrimental effect on fuel cell performance. The use of bimodal fiber structures as gas diffusion layers in the fuel cell, instead of the typically used unimodal ones, made no difference at an average or high porosity, but proved to be very beneficial at low porosity, i.e., for highly compressed fuel cell diffusion media.

I. INTRODUCTION

In a proton exchange membrane fuel cell (PEMFC), the reactant hydrogen and oxygen gases must be efficiently dispersed to the catalytic sites of the electrodes. This is accomplished by means of the two gas diffusion layers (GDL), one in the anode and one in the cathode side of the cell, which are typically made of a fibrous material, such as carbon paper or woven carbon cloth, and serve also the transport of electric current through the cell. The transport and structural properties of these layers are obviously critical to the fuel cell performance. Such properties include the fiber diameter, porosity, percolation threshold, effective diffusivity, effective thermal and electrical conductivity, and viscous permeability. Gas diffusion and catalyst layers of bimodal (two-diameter) fiber size distribution were employed in a recent experimental PEM fuel cell study, with significant beneficial effects on fuel cell performance.[1] These effects were attributed by the authors to the increased catalytic activity of the employed materials, yet they could be due in part to the different transport and structural properties arising from the bimodal distribution of fiber diameters. Furthermore, as reported in another recent experimental PEMFC study,[2] transition regime diffusion may dominate mass transport in the gas diffusion layers for a certain range of operating conditions and mean pore size. Nonetheless, practically all PEMFC modeling studies assume conditions of molecular (bulk) diffusion in both the gas diffusion layers and the catalyst zones. The objective of this work is to calculate reliable values for the transport and structural properties of bimodal fiber structures in all diffusion regimes, by properly upgrading and applying the experimentally validated numerical techniques by Tomadakis and co-workers,[3-6] and use the derived values in a two-dimensional, non-isothermal PEM fuel cell model – an upgraded version of an earlier isothermal model by Tomadakis[7-8] – to assess their effect on fuel cell performance.

II. Theoretical Principles and Methods

A. Diffusion Regimes and Structural Properties

The average distance traveled by a gas molecule between successive collisions with other molecules is the continuum mean free path, given by the kinetic theory of gases as $\overline{\lambda}=(\sqrt{2}\,\pi d_m^2 n)^{-1}$, where d_m is the mean molecular diameter and n the molecular concentration, which for ideal gas conditions is given by $n=PN_A/RT$, N_A being the Avogadro number. Obviously, the mean free path is a function of molecule size, pressure, and temperature. The average distance traveled by a gas molecule between successive collisions with the pore wall is the mean intercept length, $\overline{d}$, which serves as a measure of the mean pore size. It relates to the porosity, ε, specific surface area, S, pore surface, S_p, and pore volume, V_p, through[3] $\overline{d}=4\varepsilon/S=4V_p/S_p$, where $\varepsilon=V_p/V$ and $S=S_p/V$, V being the total volume of the porous medium. For a bimodal fiber size distribution of equal fiber density for the two fiber sizes r_1 and r_2, analytical expressions reported by Tomadakis and Sotirchos[9] may be used to derive

$$\overline{d} = \frac{-2r_1(1+\beta^2)}{\ln\varepsilon\,(1+\beta)}\;;\quad \beta = \frac{r_2}{r_1} \tag{1a,b}$$

The Knudsen number is defined as the ratio of the continuum mean free path to the mean intercept length, $Kn=\overline{\lambda}/\overline{d}$, and is commonly used to parameterize the diffusion process in porous environments. For $Kn\le 0.02$, diffusion occurs in the ordinary (bulk) diffusion regime, while for $Kn\ge 100$, intermolecular collisions are rare, and Knudsen diffusion is the dominant mode of mass transport. For $0.02<Kn<100$, both intermolecular and molecule-wall collisions contribute significantly to the mass transfer resistance, i.e., transition regime conditions prevail.

B. Transport Properties

The effective diffusivity, D_e, of a fluid in a porous medium relates to the porosity, ε, and the tortuosity factor, η, through $D_e=D_p\,\varepsilon/\eta$. The reference diffusivity, D_p, is the self-diffusion coefficient of the gas species in a cylindrical pore of diameter equal to the mean intercept length of the fibrous structure ($\overline{d}$) under the same temperature and pressure as in the porous medium.[3] When bulk or Knudsen diffusion conditions prevail, the above diffusivity relation becomes $D_e^b=D^b\,\varepsilon/\eta^b$ or $D_e^K=D^K\,\varepsilon/\eta^K$, where $D^b=1/3\,\overline{\upsilon}\overline{\lambda}$ and $D^K=1/3\,\overline{\upsilon}\overline{d}$ are the limiting values of the self-diffusion coefficient in the two regimes. The mean thermal speed of the molecules, $\overline{\upsilon}$, is given by the kinetic theory as $\overline{\upsilon}=\sqrt{8RT/\pi M}$, M being the molecular weight of the gas. In the transition diffusion regime, the reference diffusivity is given from $D_p = D^b(1+Kn)^{-1} = D^K(1+Kn^{-1})^{-1}$.

The formation factor, F, of a porous medium is defined as the inverse dimensionless effective diffusivity in the bulk regime: $F = D^b/D_e^b = \eta^b/\varepsilon$. Due to the mathematical analogy between various modes of transport, the formation factor is also equal to $F = p/p_e$, where p_e stands for the effective thermal or electrical conductivity, dielectric constant, or magnetic permeability of the porous medium, and p for the corresponding bulk property value. Using electrical conduction principles, Johnson *et al.*[10] proposed estimating the viscous permeability of a porous medium from $k = \Lambda^2/8F$, where Λ is a weighted pore volume-to-surface ratio, a characteristic length related to viscous flow. For structures where the porosity decreases uniformly when the solid phase grows into the pore space, Λ is obtained from[10] $\Lambda = -\bar{r}(d\ln F/d\ln\varepsilon)^{-1}$, where $\bar{r} = \bar{d}/2$. When the relation $F(\varepsilon)$ becomes available numerically or experimentally for the porosity range of interest, and reliable values of $\bar{d}$ are also available in that range, the above equations[10] can be used for estimating the viscous permeability, k, of the medium. The method was tested by Tomadakis and Robertson[5-6] for various types of random unimodal fiber structures, and its predictions were found in very good agreement with numerous experimental measurements and analytical results of the literature. For a bimodal fiber size distribution of equal fiber density for the two fiber sizes r_1 and r_2, Eq. (1a) may be substituted into the above relations to give the dimensionless viscous permeability of such structures:

$$\frac{k}{r_1^2} = \frac{\varepsilon}{8\eta^b \ln^2 \varepsilon}\left(\frac{1+\beta^2}{1+\beta}\right)^2\left(1 - \frac{d\ln\eta^b}{d\ln\varepsilon}\right)^{-2} \qquad (2)$$

Once the effective diffusivity of the bimodal fiber structures is obtained from the modified random walk simulation techniques,[3-4] the corresponding values of the tortuosity, formation factor, and viscous permeability are readily derived from the above discussed equations.

C. PEM Fuel Cell Model

The gas diffusion layer transport properties obtained from the random walk simulations as outlined in the previous paragraph are used in a fuel cell model within Comsol Multiphysics, along with reliable literature values for the transport properties of the catalyst and membrane layers.[11-13] The mass, momentum, electronic and ionic balances of the model were developed in an earlier investigation.[7] Heat balances were now added to the model, in all seven zones of the fuel cell; in the gas flow channels, the heat balances account for convective and conductive heat transfer; in the gas diffusion layers, they account also for the ohmic heat source due to the electric current; in the catalyst layers, three more sources of heat are accounted for, associated with the resistance to ionic current and the chemical reaction and activation thermal effects; in the membrane, the heat balance accounts for convective heat transfer due to water flow, conductive heat transfer, and an ohmic heat source due to the ionic current (resistance to proton transport). The mass, momentum, and charge balances of the model are coupled to the heat balances through the temperature effects on densities, reaction rates, concentrations, and membrane electrical conductivity. Thermal effects on diffusivity and viscosity were assumed negligible due to the anticipated small temperature variations in the fuel cell.

III. Results and Discussion

A. GDL Transport Properties

Bimodal fiber structure tortuosities computed from the random walk simulations are presented in Figure 1 for both flow configurations (transverse and in-plane) and all values of investigated porosity and Knudsen number. The tortuosity results for unimodal fibers[3] are compared in these plots to those for bimodal fibers. Excellent agreement between the two sets of results is encountered at porosities greater than 0.5, for all the selected Knudsen numbers and for both transverse and in-plane flow. For porosities lower than 0.5, there is a noticeable difference in the tortuosity due to a lower percolation threshold for bimodal fibers, 0.045 versus 0.11 for unimodal fibers (determined from a separate algorithm, measuring isolated vs. total porosity). In Figure 1a (Kn =100 and transverse flow) Faley and Strieder's[14] lower bound is also shown. The bound was derived using variational principles and is meant to predict the minimum tortuosity when $\varepsilon \rightarrow 1$.[9] It is evident from Figure 1a that the computation results are in agreement with Faley and Strieder's[14] bound. In Figure 1c (Kn=0.02 and transverse flow) the experimental data from Penman[15] and Bateman *et al.*[16] are plotted and found to be in good agreement with the random walk simulation results of both unimodal and bimodal fibers. Tsai and Strieder's[17] lower bound for bulk diffusion in transverse flow through 2-d fibers is also displayed, given by $\eta^b \geq 1 - \ln \varepsilon$. The bound agrees well with the tortuosity results for ε>0.7. It also agrees with the results of Tomadakis and Sotirchos[3] for 2-d unimodal fibers.

The formation factor of the bimodal fiber structures, $F = \eta^b / \varepsilon$, can be readily derived from the corresponding porosity and bulk tortuosity values presented in Figure 1. It offers a straight-forward calculation of the dimensionless effective thermal and electrical conductivity, magnetic permeability, and dielectric constant of the structures, as discussed in section II.B.

Figure 2 shows the computed dimensionless viscous permeability for unimodal and bimodal fiber structures. The values for a porosity of 0.2 are almost identical for the two types of fiber structures in transverse flow conditions and higher by about 70% for bimodal fiber structures in in-plane flow conditions. The values for the unimodal structures at a porosity of 0.4 are higher than the values of bimodal structures by about 80% for transverse and 30% for in-plane flow conditions. At a porosity of 0.12, in the proximity of the percolation threshold for unimodal fiber structures, the dimensionless permeability for bimodal fiber structures is higher than that of the unimodal structures by a factor of about 71 for in-plane and 24 for transverse flow conditions. This suggests that bimodal fibers should be used in applications involving highly compressed or densified media, including fuel cell gas diffusion media.

B. Fuel Cell Performance

Modeling results were obtained both for unimodal and bimodal GDL fibers based on the earlier PEMFC model developed by Tomadakis[7] updated with heat balance equations in all seven zones of the fuel cell. The model was evaluated at three different porosity values: 0.7, 0.4, and 0.12. The temperature of both feed gas streams was set to 353 K, and their pressure and relative humidity to 2 atm and 25%, respectively.

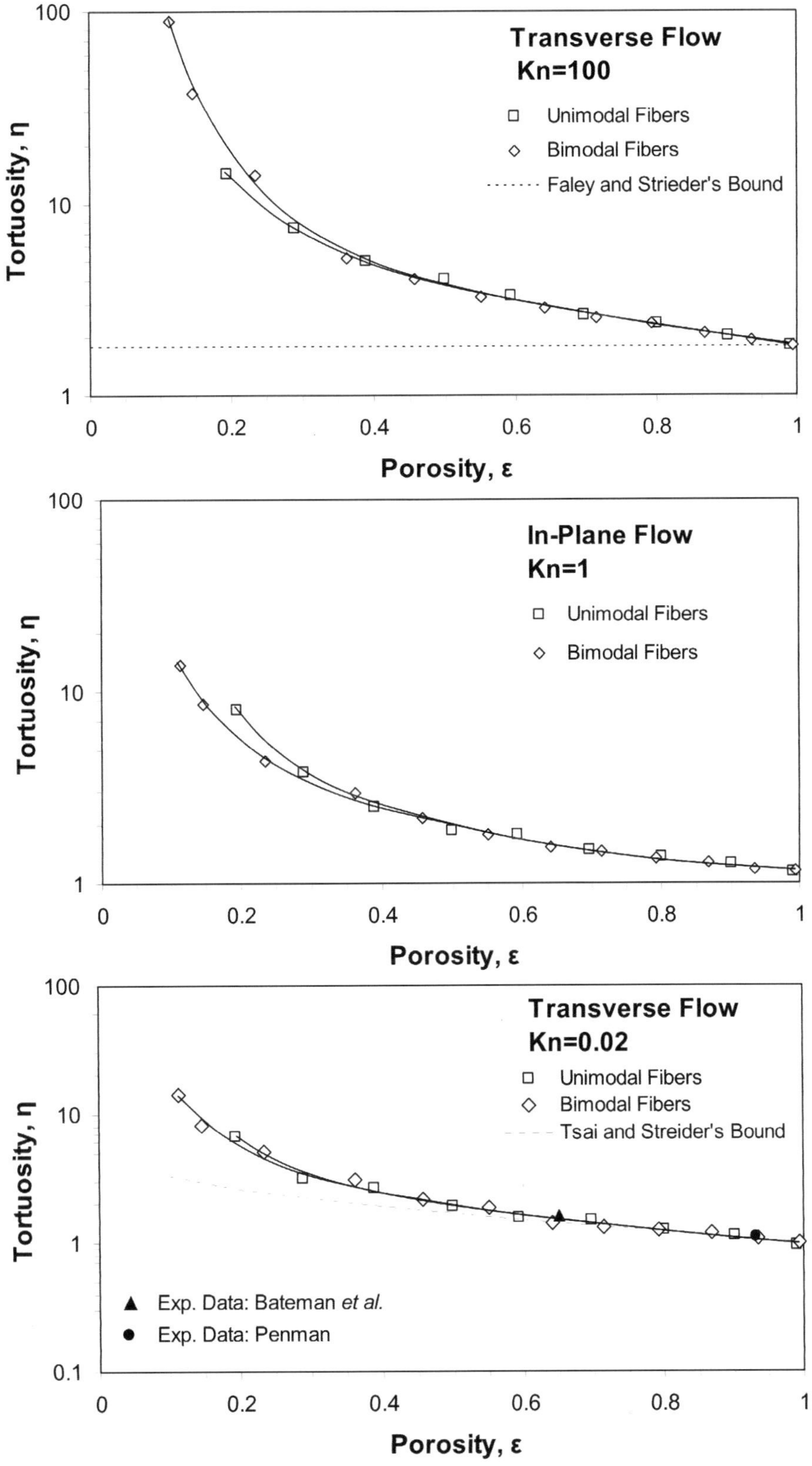

Figure 1. Unimodal vs. Bimodal Fiber Structure Tortuosity for Kn=100, 1, 0.02 and Transverse or In-Plane Flow

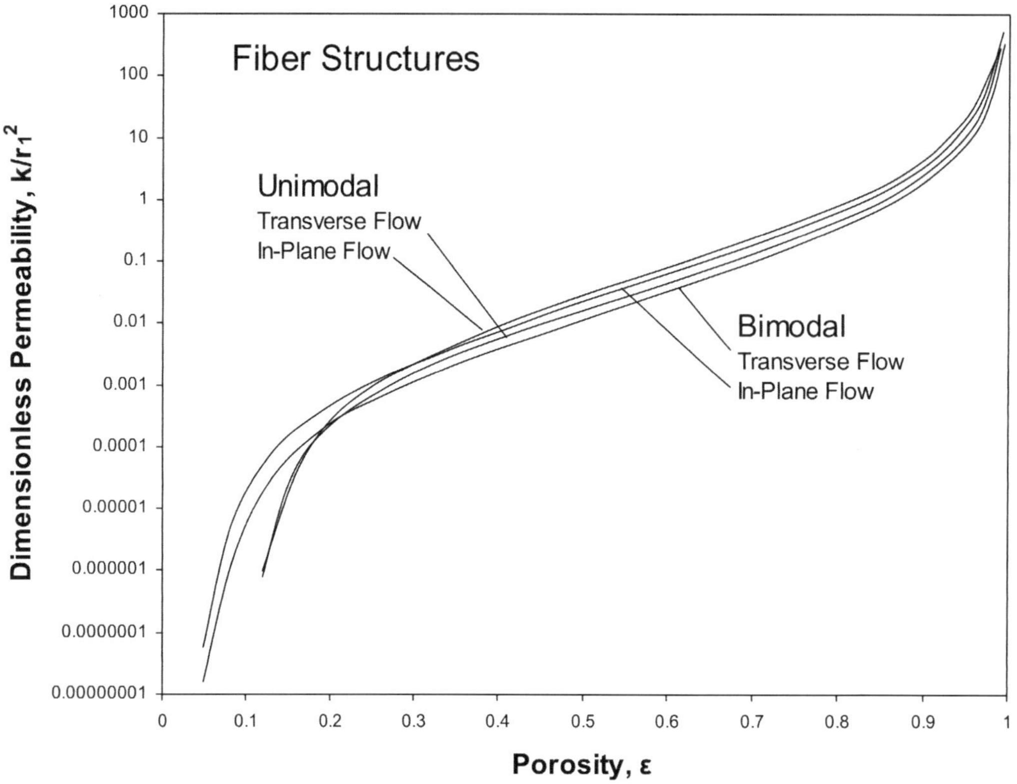

Figure 2. Viscous Permeability of Bimodal vs. Unimodal Fiber Structures

Figure 3 shows the effect of the GDL porosity on fuel cell power density, for unimodal GDL fibers. The lowest power density is obtained with the lowest porosity, 0.12, close to the percolation threshold of the media (ε_p=0.11). At this value, the effective diffusivity of the reactants through the GDL's and to the catalytic sites is extremely low, resulting in a degraded fuel cell performance. The highest power density is obtained with a GDL porosity of 0.4 and not with the highest porosity (0.7), where higher yet effective diffusivity is attained. This is attributed to the fact that when the porosity increases, the fiber occupied space decreases, thereby resulting in decreased electrical conductivity, an adverse effect that proves to be slightly stronger than the positive effect of higher diffusivity.

The maximum temperature rise in the fuel cell varies from 0 to 5.4 K, with the largest values obtained with the high porosity of 0.7. The computed values are in agreement with experimental and modeling results of the literature.[11, 18-21] For a porosity of 0.4 or 0.7, there is no difference in the fuel cell power density results obtained using unimodal or bimodal GDL fibers, due to the practically equal values of bulk tortuosity and viscous permeability of the two types of fiber structures. At a porosity of 0.12, it is found beneficial to use bimodal GDL fibers, especially at higher current densities, as is evident in Figure 4, which offers a comparison of the power density results obtained with unimodal and bimodal GDL fibers at that porosity level. As mentioned earlier, bimodal fiber structures have a lower percolation threshold (0.045 vs. 0.11), so there is more activity in the cell at this lower porosity value. Bimodal fiber structures would therefore be beneficial as GDL diffusion media in highly compressed fuel cells, or in other applications involving compressed beds of fibers.

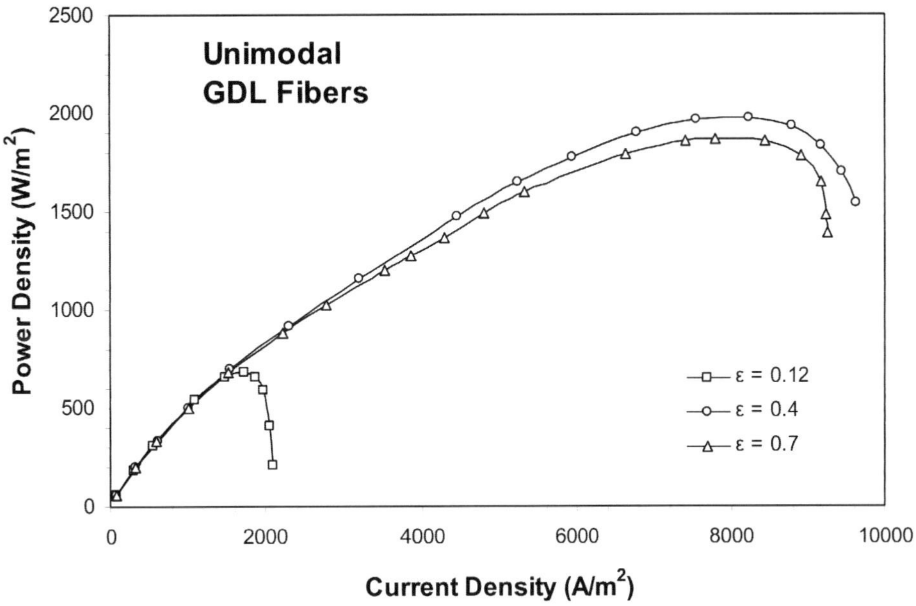

Figure 3. GDL porosity effect on fuel cell performance

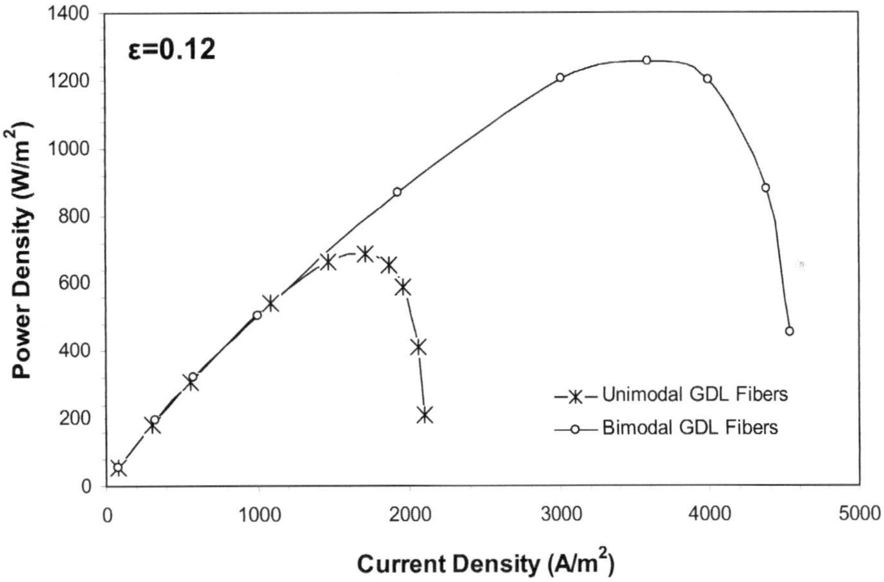

Figure 4. GDL fiber size distribution effect on fuel cell performance

A comprehensive analysis of the values of all model parameters that influence the mode of diffusion (diffusion regime) in the PEMFC gas diffusion layers reveals that bulk diffusion dominates GDL mass transport for all porosities and GDL operating conditions discussed so far. Transition regime diffusion can be achieved at ε=0.12 by reducing the fiber radius to nanofiber levels (r_f = 0.1 μm) or by lowering the pressure to a vacuum of 0.04 atm. Nonetheless, such a very low pressure results in extremely low values of power density, due to its adverse effect on the reactant gas concentrations. Figure 5 presents the power density curves obtained for bimodal GDL fibers with flow in the bulk or transition regime, the latter accomplished by setting the fiber radius to 0.1 μm. As deduced from this plot, the reduction of the fiber size by a factor of 50 has a very detrimental effect on the performance of the fuel cell, primarily due to the reduction of the GDL viscous permeability by a factor of 50^2=2500 (since k/r_f^2 remains constant at fixed porosity), which limits drastically the activity in the cell.

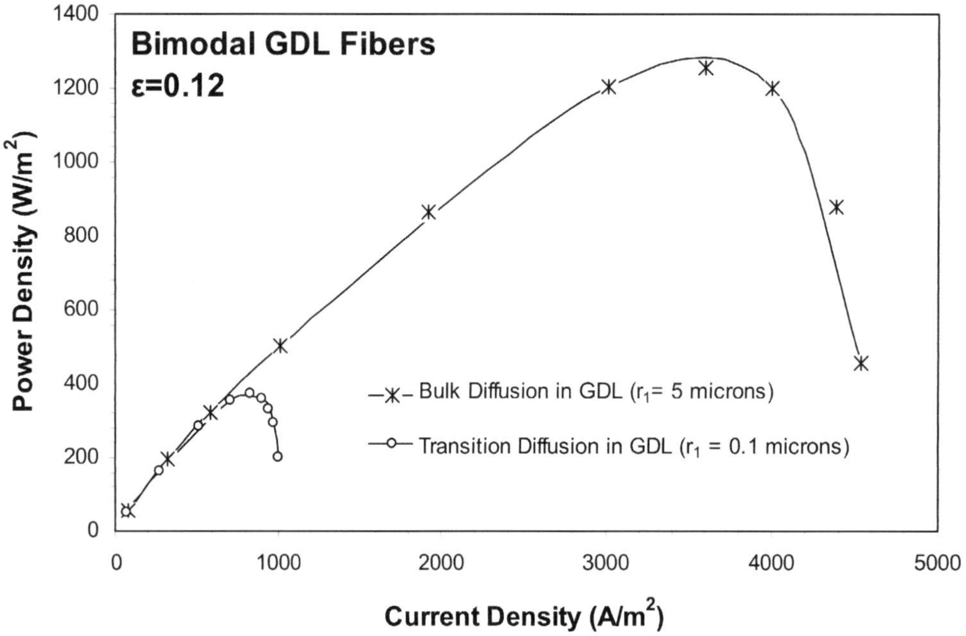

Figure 5. GDL diffusion regime effect on fuel cell performance

REFERENCES

[1]Chang B, Lu Y, Tatarchuk B (2006) *Chem. Eng. J.*, 115, 195-202.

[2]Williams M, Begg E, Bonville L, Kunz H, Fenton J (2004) *J. Electroch. Soc.*, 151, A1173-80.

[3]Tomadakis MM and Sotirchos SV (1993) *AIChE Journal*, 39, 397-412.

[4]Tomadakis MM and Sotirchos SV (1993) *J. Chem. Phys.* 98, 616-626.

[5]Tomadakis MM and Robertson TJ (2005) *J. Compos. Mater.*, 39, 163-188.

[6]Tomadakis MM and Robertson TJ (2005) *J. Chem. Phys.*, 122, 579-588.

[7]Tomadakis MM (2006) *Hydrogen as Fuel for Aircraft and Aerospace Applications Research Project, Year-2 Final Report to NASA Glenn Research Center.*

[8]Tomadakis MM and Pentas KH (2006) *AIChE Spring National Meeting CD*, pp. 1-6.

[9]Tomadakis MM and Sotirchos SV (1991) *AIChE J.*, 37, 74-86.

[10]Johnson DL, Koplik J, Schwartz LM (1986) *Phys. Rev. Lett.*, 57, 2564-2567.

[11]Rowe A and Li X (2001) *J. Power Sources*, 102, 82-96.

[12]Sivertsen BR and Djilali N (2005) *J. Power Sources*, 141, 65–78.

[13]Ju H, Meng H, Wang CO (2005) *Int. J. Heat Mass Trans.*, 48, 1303-1315.

[14]Faley TL and Strieder W (1987) *J. Appl. Phys.*, 62, 4394-4397.

[15]Penman HL (1940) *J. Agric. Sci.*, 30, 437-462.

[16]Bateman BR, Way JD, Larson KM (1984) *Sep. Sci. Tech.*, 19, 21-32.

[17]Tsai DS and Strieder W (1986) *Chem. Eng. Comm.* 40, 207-218.

[18]Djilali N and Lu D (2002) *Int. J. Therm. Sci.* 41, 29–40.

[19]Shimpalee S, Greenway S, Spuckler D, Van Zee, J (2004) *J. Power Sources*, 135, 79–87.

[20]Ziegler C, Tranitz M, Schumacher JO (2005) *FEMLAB Conference CD.*

[21]Ma Z, Jeter SM, Abdel-Khalik SI (2003) *Int. J. Hydrogen Energy*, 28, 85-97.

33b Gas Diffusion Layers in Direct Methanol Fuel Cells (DMFC)

Virendra K. Mathur, Dept. of Chem. Eng., University of New Hampshire, 260 Kingsbury Hall, Durham, NH 03824 and Jason Morgan, Chemical Engineering, Dept. of Chem. Eng., University of New Hampshire, 260 Kingsbury Hall, Durham, NH 03824.

An automated bench-scale fuel cell experimental set-up (PEMFC/DMFC) has been designed and fabricated. The experimental unit is capable of controlling the mass flow rates of gases and liquids, humidity of the inlet gases, as well as the temperature and pressure of the cell in a safe and efficient manner.

The optimum cell temperature and methanol solution concentration are found to be 75°and 4 molar, respectively. The methanol flow rate is found to be 3 mL/min with catalyst loadings on both fabric and paper GDLs of about 2.5 mg/cm2. A study into the effects of PTFE (Teflon") content in both the substrate and microporous layer (MPL), as well as the MPL loading is also conducted. It is found that the anode GDL for a DMFC system should have high PTFE content in the substrate and be coated with a thick MPL with a low PTFE content. From these results a new GDL system, consisting of a paper anode GDL modified with a microporous sublayer and an unmodified paper cathode GDL is developed. The result of this development work is a GDL system which can be manufactured continuously providing improved performance compared to a commercial reference.

33c Gas and Water Management in a Dead-Ended Operating PEM Fuel Cell

Michael S. Pien, Marvin Warshay, Steven Lis, and Radha Jalan. ElectroChem, Inc, 400 W. Cummings Park, Woburn, MA 01801

A dead-ended operating Proton Exchange Membrane (PEM) Fuel Cell is favorable for high efficiency of reactant gas utilization and self-humidification and permits the simplification of components in a balance of plant design. However, special gas and water management is required for a dead-ended operating PEM fuel cell to ensure stable and reliable performance.

For conventional PEM fuel cells, the gas and water management is usually addressed by applying a forced flow-through-gas to create a significant pressure difference between the inlet and outlet of each channel that provides uniform gas distribution through the cell and strong gas flow to remove product water. It usually requires very narrow flow channels to create the needed flow rate and pressure. In some other cases, fuel cell operators will "burp the stack" causing a sudden release of gas in the hope of removing the water droplets.

In a dead-ended operating PEM fuel cell where pure H_2 and O_2 are applied, it creates a very different scenario of operation. When pure O_2 is used, flow-through-gas operation is not applicable. A rapid circulation of pure O_2 is also not desirable because of the potential fire hazards associated with fast moving mechanical components in a pure oxygen atmosphere. ElectroChem, Inc. has developed an innovative PEM Fuel Cell with an open channel structure which provides a mechanism of gas and water management for the dead-ended operating PEM fuel cell.

I. Demonstration of stable operation at extremely low excess oxygen flow

The fuel cell demonstrated a stable operation at extremely low oxygen flow rates, i.e., at extremely low percentages of excess oxygen (over stoichiometric value). The cell required a thirtieth of the excess oxygen of that the conventional cell requires to maintain stable operation. The demonstration of the fuel cell with stable operation at extremely low excess oxygen flow rates is indicated in the test results seen in Figure 1. To reveal the effect of low oxygen excess flow rate on the cell performance, the tests were taken with two consecutive runs with a difference in the oxygen excess flow rate.

Figure 1. Effect of oxygen excess flow rates on the cell performance

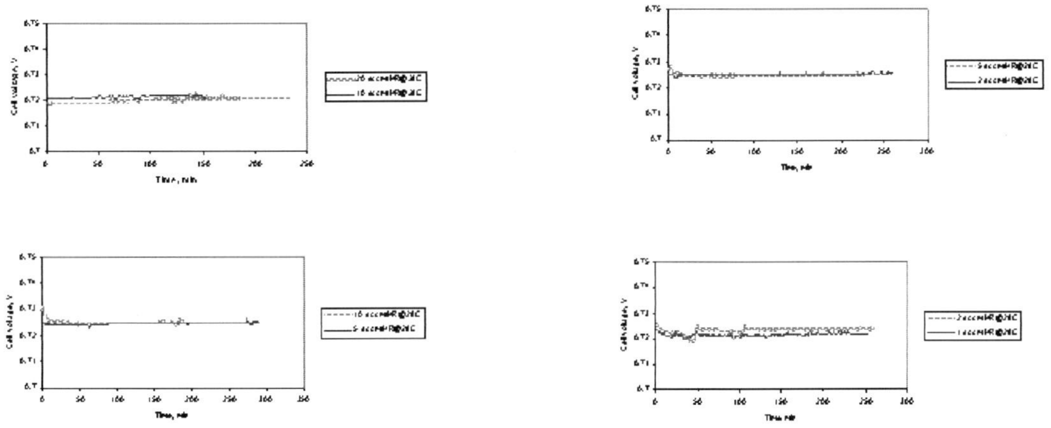

II. No effects of change in orientation on cell performance

The fuel cell operated under four different orientations, 0°, 90°, 180°, and 270° with respect to the location of the cell's oxygen exit. At 180°, the oxygen exit is located at the top of the cell in which the production water is required to move upward against the gravity to leave the cell. As shown in Figure 2, there is no significant effect on product water removal when the cell is at 180° orientation. These test results strongly implied "against gravity" product water removal capability of the cell.

Figure 2. Effect of orientation on the cell performance

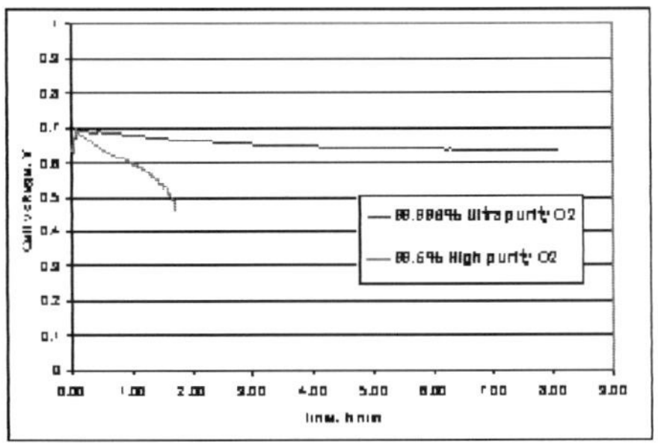

Cell temperature: 75°C

Cell pressure: 30 psig

O2 excess flow rate: 2 cc/min

H2 excess flow rate: 2 cc/min

III. Effect of impurity in the reactant gas on cell performance

In a dead-ended operating PEM fuel cell, accumulation of inert gas can reduce the partial pressure of the reactant gas and affect the cell performance. To investigate the impurity effect, a ultra-high-purity oxygen (99.996% O_2) is used to compare with a high-purity oxygen (estimate 99.5% O_2). As shown in Figure 3. a dramatic drop in the cell performance was caused by the accumulation of inert gas while operating at a dead-ended condition.

Figure 3. The effect of inert gas on a dead-ended operating PEM fuel cell

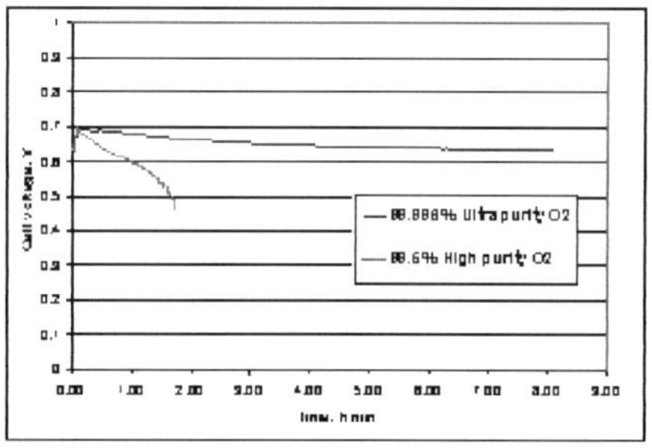

Cell temperature: 75°C

Cell pressure: 30 psig

Current density: 400 mA/cm^2

The open channel cell demonstrated a different gas and water management mechanism which permitted the PEM fuel cell to operate at almost no excess oxygen and hydrogen flow condition and transport product water against gravity. However, the purity of the reactant gas will determine whether the cell can be operated at near dead-ended or complete dead-ended conditions.

Acknowledgments

This work was funded by the National Aeronautics and Space Administration under Contract NNJ06JD71C. Helpful discussions with the NASA teams are gratefully acknowledged.

Web Page: www.fuelcell.com

The Rate of Carbonate Formation in a Direct Carbon Fuel Cell

Research performed by:
Charles W. Kling
Dept. of Chemical Engineering, West Virginia University

Mentored by:
Dr. John W. Zondlo
Dept. of Chemical Engineering, West Virginia University

Research supported by:
Summer 2008 NASA Undergraduate Research Fellowship
West Virginia Space Grant Consortium

ABSTRACT:

The production of carbonate (CO_3^{2-}) in a Direct Carbon Fuel Cell (DCFC) was investigated. The DCFC is a technology that can produce electricity from solid carbon fuel sources with no moving parts and, theoretically, at much higher efficiencies than conventional coal-fired power plants. Carbonate is produced in the electrolyte through an undesirable side reaction and poisons the cell at higher concentrations. A bench-scale DCFC utilizing a 600°C molten sodium hydroxide (NaOH) electrolyte was operated and tested to determine the rate of carbonate formation. A wet-chemistry procedure was developed to measure the mass fraction of sodium carbonate (Na_2CO_3) in a solidified sample of the electrolyte. This mass fraction was used to calculate the concentration of the carbonate ion in the molten electrolyte. Multiple electrolyte samples were analyzed, and the concentration of carbonate was plotted as a function of fuel cell operation.

It was found that the formation of carbonate varied linearly with the energy produced by the fuel cell. Specifically, 0.67 mol of carbonate are produced for every 100 kJ of energy produced.

These data represent a significant step in DCFC research, as the formation of carbonate, while known to be undesirable, has never been charted as a function of fuel cell operation. Attempts to reduce the rate of carbonate formation can now be made and compared against this base rate to judge the relative success or failure of these attempts.

Additionally, in an effort to reduce the formation of carbonate, doping the DCFC electrolyte with magnesium oxide (MgO) was also investigated. It was found that there was a time lag before carbonate was produced; however, once carbonate began forming, the rate was unchanged compared to the un-doped electrolyte. Hence, the MgO delayed, but did not reduce, the formation of carbonate in the electrolyte.

INTRODUCTION:

The formation of carbonate (CO_3^{2-}) in a Direct Carbon Fuel Cell (DCFC) was investigated. The DCFC is a technology that produces electricity from solid carbon fuel sources electrochemically, without the energy losses associated with combustion or mechanical electricity generating equipment. Theoretically the DCFC can achieve efficiencies over twice those of conventional coal-fired power plants.

While there are many perceived advantages to using direct carbon fuel cells, there are also several technological hurdles which must be cleared before the technology becomes viable for large scale electricity production. One of the most significant hurdles is the occurrence of an undesirable side reaction in the electrolyte. Carbon dioxide (CO_2) produced at the anode reacts with hydroxide ions (OH^-), which are present in the electrolyte, to produce carbonate (CO_3^{2-}). The mechanism for this reaction is not fully understood, but one likely mechanism has been proposed by J. Goret and B. Tremillion [1]. This mechanism consists of a chemical process, shown in Equation 1, and an electrochemical process shown in Equation 2. The electrochemical process consists of two steps shown in Equations 3 and 4, with the reaction in Equation 4 being the rate determining step [1].

$$CO_2 + 2OH^- \leftrightarrow 2CO_3^{2-} + H_2O \qquad (1)$$
$$C + 6OH^- \rightarrow CO_3^{2-} + 3H_2O + 4e^- \qquad (2)$$
$$6OH^- \leftrightarrow 3O^{2-} + 3H_2O \qquad (3)$$
$$C + 3O^{2-} \rightarrow CO_3^{2-} + 4e^- \qquad (4)$$

For this work, a bench-scale DCFC utilizing a 600°C molten sodium hydroxide (NaOH) electrolyte was operated and tested to determine the rate of carbonate formation. A wet-chemistry procedure was developed to measure the mass fraction of sodium carbonate (Na_2CO_3) in a solidified sample of the electrolyte. This mass fraction was used to calculate the concentration of the carbonate ion in the molten electrolyte. Multiple electrolyte samples were analyzed, and the concentration of carbonate was plotted as a function of the energy produced by the fuel cell.

Additionally, in an attempt to reduce the formation of carbonate, doping the sodium hydroxide electrolyte with magnesium oxide (MgO) was also investigated. Doped electrolyte samples were collected, analyzed, and compared with the base carbonate formation rate described herein.

BACKGROUND:

Theoretically, the DCFC has the potential to make electricity from solid carbon-based fuel sources, such as coal, with an efficiency of around 80%, which is more than twice the efficiency of conventional coal-burning power plants (approx. 35%). In a DCFC, electricity is produced from solid carbon electrochemically, utilizing the same overall reaction as combustion, but without the parasitic energy losses associated with combustion. Additionally, since electricity is generated directly from the fuel cell, no energy is lost through inefficiencies in mechanical equipment, such as turbines, used in traditional power generation. Other benefits of the DCFC include the evolution of a concentrated waste stream of carbon dioxide

(CO_2) which should allow for easier carbon sequestration when that technology becomes viable. Electricity is produced in the DCFC by the reactions shown in Equations 5, 6, and 7.

Anode Reaction

$$C + 4OH^- \rightarrow CO_2 + 2H_2O + 4e^- \qquad (5)$$

Cathode Reaction

$$4e^- + O_2 + 2H_2O \rightarrow 4OH^- \qquad (6)$$

Overall Reaction

$$C + O_2 \rightarrow CO_2 \qquad (7)$$

Early research on DCFC technology was performed by Dr. William Jacques in 1896 [4] and more recently at Scientific Applications & Research Associates (SARA) Inc. [5] and West Virginia University (WVU) [2].

A diagram of the bench-scale DCFC utilized for this work is shown in Figure 1. The fuel cell consists of a cylindrical nickel (Ni) pot which contains a molten metal-hydroxide electrolyte, typically sodium hydroxide (NaOH) at 600°C, with a graphite fuel rod suspended in the electrolyte. The nickel pot plays the role of the cathode, while the carbon rod itself is the anode. Oxygen (O_2) and water (H_2O) are supplied to the cell in the form of humidified air through a "spider" bubbler located at the bottom of the cell. Electrons from the anode power an electric load and then return to the cathode to complete an electrical circuit and the redox reaction.

The fuel cell utilized for this work is supported by a fuel cell stand, donated by the Department of Energy (DOE), which supplies the infrastructure necessary to operate and monitor the cell. This infrastructure includes a heater, gas regulators, various sensors, a load cell, and a data logger as well as necessary safety containment equipment. The load cell draws a constant current load on the fuel cell, allowing the voltage to fluctuate based on the power produced by the cell. During fuel cell operation, the data logger records cell parameters such as air flow rate, cell current, cell voltage, and various temperatures as a function of time.

EXPERIMENT:

Base Rate Experiment:

Before any attempt could be made to reduce the rate of carbonate (CO_3^{2-}) formation in the sodium hydroxide (NaOH) electrolyte, a base rate needed to be determined for proposes of comparison. A search of the relevant literature was conducted and no data were found on the rate of carbonate formation. Goret and Tremillon proposed humidifying the electrolyte as one way to reduce the rate of carbonate formation, but did not quantitatively define this rate or how much the rate could be reduced by electrolyte humidification [1].

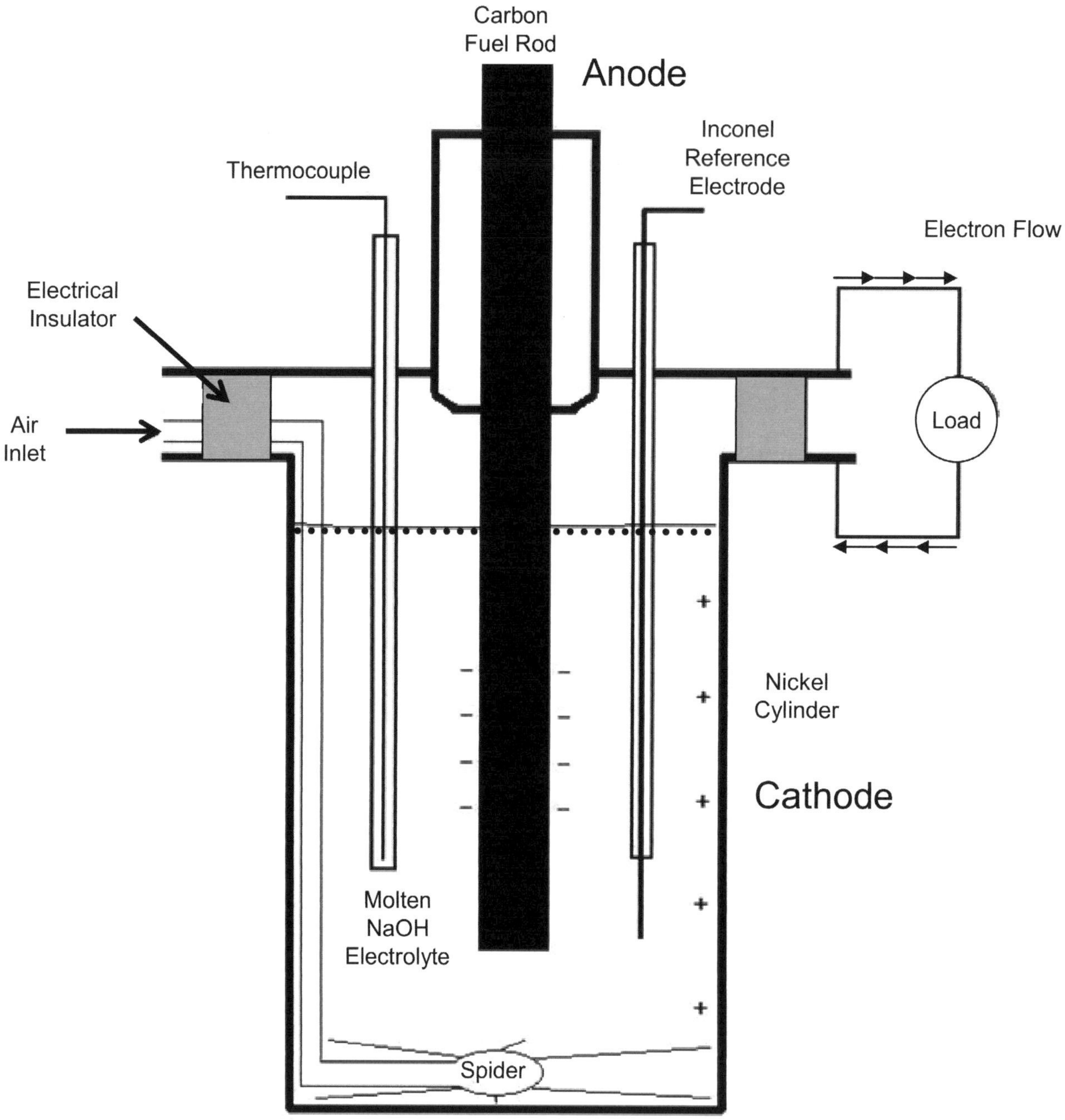

Figure 1: Diagram of Bench-Scale Direct Carbon Fuel Cell

In an analytical chemistry text titled Quantitative Chemical Analysis, in a section on "Preparation of Solutions to be Standardized", a technique is described to purify sodium hydroxide contaminated with carbonate due to exposure to carbon dioxide from the air [3]. By modifying this technique, a procedure to measure the mass fraction of sodium carbonate (Na_2CO_3) in a solidified sample of the fuel cell electrolyte was developed. The measured mass fraction was then used to calculate the concentration of carbonate ions in the molten electrolyte at the time the sample was taken.

Typically, four or five samples of the molten electrolyte were withdrawn over a five day period of fuel cell operation. To withdraw a sample, the anode assembly, including the graphite rod, was temporarily removed from the cell and a stainless steel crucible was used to retrieve approximately 50 g of the molten electrolyte solution. The sample which is hygroscopic was allowed to cool and then sealed against moisture for storage. It should be noted that there are significant safety concerns involved when handling a 600°C molten hydroxide solution and appropriate safety procedures must be observed during cell operation and especially when sampling the electrolyte.

To analyze the solidified sample, it was first weighed and then dissolved in deionized water. After the sample dissolved, sodium hydroxide was added to bring the solution to a concentration of 12 mol NaOH / liter (12 M). Since the sample itself was primarily sodium hydroxide, care was taken to keep the solution close to the desired 12 M concentration. Sodium carbonate, which is soluble in water, is not soluble in strong solutions (over 10 M) of sodium hydroxide and will precipitate out of solution as a solid white powder. After the sodium carbonate precipitate appeared in the 12 M solution, the solution was filtered using a standard vacuum filtration apparatus. The filter cone and paper, containing the sodium carbonate, were then dried and weighed. The weight of the precipitated sodium carbonate and the original weight of the sample were compared to give a mass fraction of sodium carbonate in the sample. This mass fraction was then used to calculate the molar concentration of the carbonate ion in the molten electrolyte.

The data logger connected to the fuel cell stand records the cell voltage, current, and other parameters at one minute intervals during fuel cell operation. This information was used to calculate the cumulative energy produced by the fuel cell at the time each sample was withdrawn. The amount of carbonate produced was then plotted against the energy produced and a rate of carbonate formation was determined.

Magnesium Oxide Doped Electrolyte Experiment:

In the earliest DCFC research, Jacques suggested that the carbonate contamination of the electrolyte may be reduced "by adding a small percentage of oxide of magnesium" to the electrolyte [4]. He claimed that with this addition, his fuel cell was able to run for many months without becoming clogged with carbonate. However, the "small percentage" that Jacques mentioned is not quantified in any literature and his results have yet to be reproduced in modern times.

After a base rate of carbonate formation was determined, the addition of magnesium oxide (MgO) and its affect on carbonate production was also investigated. In deference to the small percentage of magnesium oxide suggested by Jacques, the DFCF described herein was operated with 2.5 wt % of MgO added to the NaOH electrolyte. The electrolyte was sampled during fuel cell operation and the samples were analyzed using the same procedure developed to determine the base rate. The carbonate concentrations were then plotted as a function of the energy produced by the fuel cell and compared to the base rate.

RESULTS:

Base Rate Results:

To determine the base carbonate formation rate, twelve samples were obtained over the course of three successful fuel cell runs. These samples were analyzed and plotted against the energy produced by the cell. As shown in Figure 2, carbonate is produced at a rate that is linearly proportional to the energy produced by the cell. Specifically, 0.67 mol of carbonate are produced for every 100 kJ of energy produced. These data establish a good baseline with which to compare carbonate production rates when attempts are made to reduce carbonate formation in a direct carbon fuel cell.

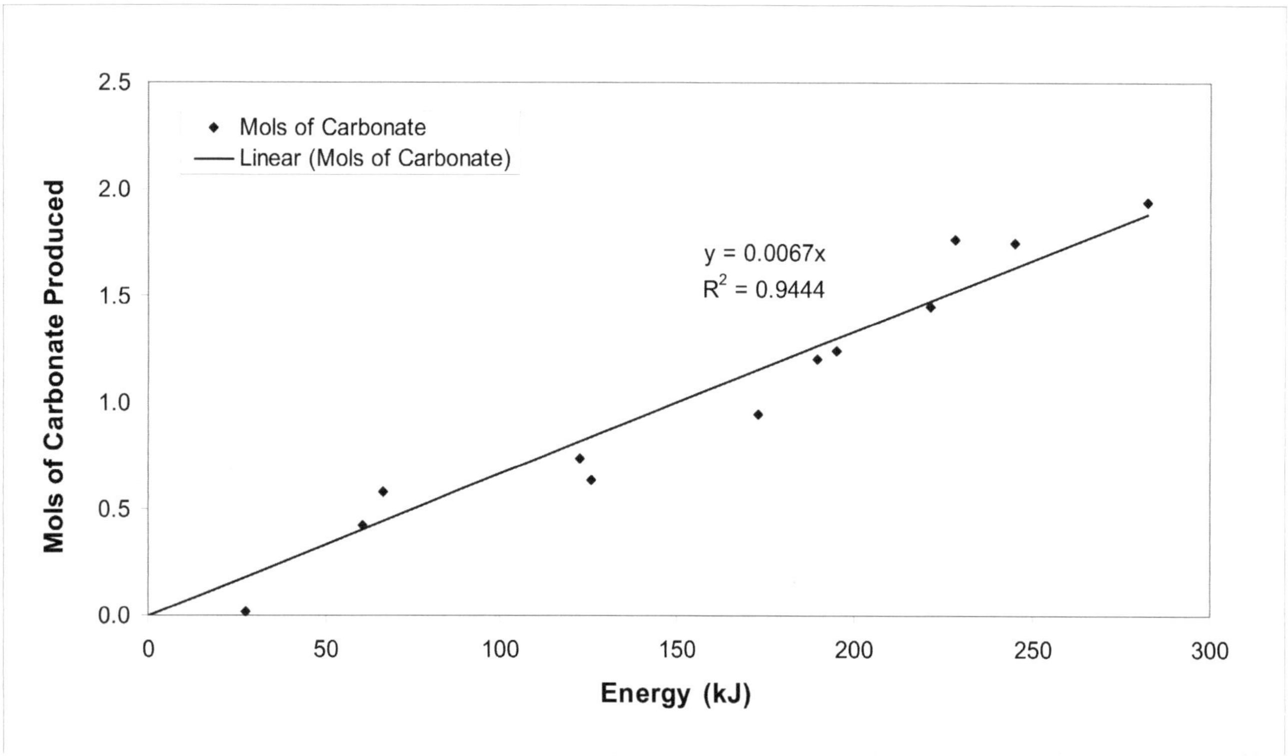

Figure 2: Base Rate of Carbonate Formation in the Direct Carbon Fuel Cell

Magnesium Oxide Doped Electrolyte Results:

In an attempt to reduce the carbonate formation rate, the sodium hydroxide electrolyte was doped with 2.5 wt % magnesium oxide (MgO). Samples of the electrolyte were taken and analyzed in the matter previously described. It was found that there was a lag time before carbonate was detected, however after carbonate formation began, the measured rate was unchanged from the base rate. Figure 3 shows the carbonate concentrations in the doped electrolyte in comparison to those of the un-doped electrolyte. It should be noted that the slopes of the two trend lines are nearly identical indicating that the formation rate is unchanged.

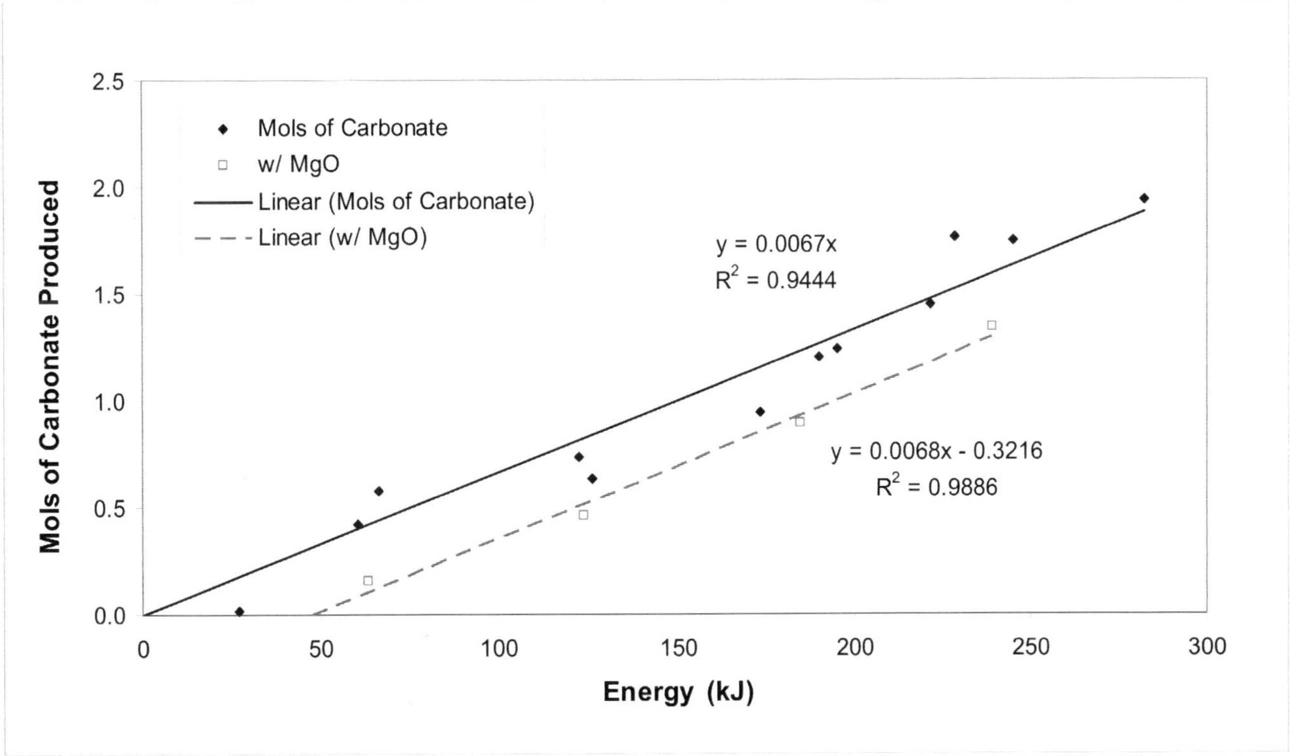

Figure 3: Carbonate Formation Rate with MgO Doped Electrolyte

DISCUSSION:

As can be seen from the reactions shown in Equations 1 and 3, an increased water concentration in the electrolyte should serve to reduce the rate of carbonate formation by driving these two equilibrium reactions to the left. For the work presented here, the water concentration in the electrolyte was held constant to eliminate the effect of water concentration variance on the carbonate formation rate.

In addition to electrolyte water concentration, the performance of the fuel cell is also sensitive to other operating parameters such as air flow rate, current load, and various temperatures. Some optimum parameter values, described herein, were narrowed down by previous work conducted on a similar DCFC at WVU [2]. Additionally, before gathering the data presented here, multiple experimental fuel cell runs were conducted to determine optimum parameters for this experiment. For this work, care was taken to hold all operating parameters constant for the multiple fuel cell runs conducted over the many weeks of fuel cell operation required for this experiment. It should be noted that these operating parameters need to be held constant during future experiments to ensure that any observed changes are from the experiment and not from a parameter change.

For this experiment a fuel cell constructed from an 8.5 inch long, 3 inch ID Ni-200 cylinder was used. A 0.50 inch diameter graphite rod was used for the carbon fuel source. 1300 grams of ACS certified NaOH pellets were melted for the base electrolyte. The fuel cell was maintained at 600°C by an external heater. Compressed breathable air regulated to 0.20 Standard Liters per Minute (SLPM) was used to supply oxygen (O_2) to the cell. To supply water (H_2O) to the cell, the air feed was humidified to saturation at 80°C. This was accomplished by bubbling the air through a tank of distilled water which was kept at 80°C. The humidified air line leaving the water tank was kept at 90°C to prevent water from condensing out of the humidified air. The humidified air line running into the fuel cell was heated to 600°C to avoid cooling the electrolyte. During normal fuel cell operation, a constant current load of 1.0 ampere was drawn from the cell by the load cell.

When the electrolyte was doped with a small percentage of magnesium oxide, it was found that there was a delay before carbonate was detected, however after carbonate formation began, the rate was unchanged from the base rate.

It could be theorized that the magnesium oxide additive reacted preferentially with carbon dioxide, which is produced at the anode, possibly forming magnesium carbonate ($MgCO_3$), causing a delay in carbonate formation. After the magnesium oxide was consumed, carbonate formation resumed at the base rate. Further experiments and analysis will be needed to confirm or disprove this theory.

CONCLUSIONS:

The formation of carbonate, which can poison a DCFC, is a major hurdle which needs to be overcome before this technology can be efficiently used for large scale power generation. The first step to reduce the formation of carbonate was to determine the formation rate. In the work presented here, the carbonate formation rate has been plotted for one set of operating parameters. Future attempts to reduce the carbonate production rate can be compared to this base rate to judge the relative success or failure of these attempts.

In early DCFC research, Jacques suggested that small percentages of magnesium oxide could indefinitely prevent the build up of carbonate in the electrolyte [4]. The work presented here has disproved this theory, at least for one particular percentage of magnesium oxide.

While it may never be possible to completely prevent the undesirable, carbonate forming, reactions from occurring, the rates of these reactions will need to be reduced significantly before DCFC technology can be used as a cost effective power source. The work presented here is but one small step towards the viability of this intriguing technology.

The work contain here has revealed several avenues for future study. Only a single percentage of magnesium oxide was investigated for this work. It is possible that future studies may reveal that other percentages give better results. However, reducing the carbonate formation rate by humidifying the electrolyte appears more promising. It is recommended that future work to reduce the rate of carbonate formation be conducted in this area. Other topics for future study include investigating more suitable materials of construction for electrolyte containment to address corrosions issues experienced with the nickel-200 cylinder utilized for this work.

ACKNOWLEDGEMENTS:

As with most ventures of any size, this experimental work could not have taken place without the support and encouragement of several individuals and organizations. I would like to thank the following people and organizations for their contributions to this research.

The NASA West Virginia Space Grant Consortium managed by Candy Cordwell provided funding for this project. Dr. John Zondlo mentored this project with infinite patience and wisdom. My wife, Anne Kling, put up with my late night and weekend trips to the lab. Greg Hackett conducted significant DCFC research at WVU and brought this project to the point where I picked it up. The United States Department of Energy (DOE) donated the fuel cell stand which provided the infrastructure for this experiment. Dr. Chunchuan Xu kept an eye on the fuel cell when I was not in the lab. Jim Hall provided ongoing machine shop support and fabricated an entirely new cell to replace one which became corroded beyond use. Adrienne McGraw ran the Scanning Electron Microscope (SEM) which helped diagnose a corrosion issue with the Ni cell. Liviu Magean provided ongoing lab support and suggested a purity test for the precipitated Na_2CO_3.

WORKS CITED:

1. D. Cao, Y. Sun, G. Wang. Direct carbon fuel cell: Fundamentals and recent developments. Journal of Power Sources. Volume 167. pp 250-257. 2007.

2. G. Hackett, J. Zondlo. Evaluation of Carbon Materials for Use in a Direct Carbon Fuel Cell. M.S. Thesis. West Virginia University. 8 May 2006.

3. L. Hamilton, S. Simpson. Quantitative Chemical Analysis. p.156. 12th ed. The Macmillan Co. New York. 1967.

4. W. Jacques. Coal Battery. Harper's New Monthly Magazine. circa 1896.

5. S. Zecevic, E. Patton, P. Parhami. Electrochemistry of Direct Carbon Fuel Cell Based on Metal Hydroxide Electrolyte. Scientific Applications & Research Associates, Inc. 30 July 2003.

33e Conductive Solid Polymer Electrolyte for Energy Storage Applications

Marie C. Kane and Ming Au. Materials Science and Technology, Savannah River National Laboratory, 773-A, Savannah River Site, Aiken, SC 29808

Commercial hybrid electric vehicles (HEV) and fuel cell vehicles (FCV) serve as a means to reduce the nation's dependence on oil. Current electric vehicles use relatively heavy metal hydride-nickel (MH-Ni) rechargeable batteries. Li-ion rechargeable batteries have been developed extensively as the replacement; however, the high cost and safety concerns are still issues to be resolved before large-scale production. Before fuel cell vehicles can become a commercial reality, significant research and development is needed to reduce material costs and improve fuel cell efficiencies. However, new battery systems and fuel cells are being developed to replace those currently in use. In this study, we propose a new highly conductive solid polymer electrolyte that has several advantages over current technologies. The polymer electrolyte system consists of mainly a synthetic water-soluble matrix with small co-polymer acidic domains. Low molecular weight monomers were infused into the polymer matrix to raise the amorphous content. The results of various polymer syntheses optimized for the energy storage system of interest to SRNL and room temperature conductivity measurements and performance assessment will be presented.

Three Dimensional Modelling of PEM Fuel Cells: Heterogeneity Analysis of Pore Structure on Fuel Cell Hydrodynamics

Navraj Hanspal, University of Manchester, UK
Joseph Dawes, University of Manchester, UK
Ali Turan, University of Manchester, UK

Abstract

In this work a three-dimensional mathematical model is presented for Polymer Electrolyte Membrane Fuel Cells (PEMFCs) that can be used for analysing the effects of pore structure heterogeneities on fuel cell hydrodynamics and in turn on fuel cell performance. Water management and flooding are critical issues in the design of (PEMFCs) which can greatly depend on the heterogeneous cathode and anode pore structures. Current three-dimensional PEMFC models either do not consider these effects or tend to over simplify the problem by making the questionable isotropic assumption regarding permeability structure. To overcome these deficiencies, an extended network permeability model has been developed in which a network of pores and capillaries is representative of the random porous media structure within the fuel cell. The network model has been developed in MATLAB and is capable of simulating heterogeneous pore distributions within the PEMFCs. To model the fuel cell flow dynamics the free flows in the gas channels have been modeled by the Navier-Stokes equations for laminar regimes whereas the flows in very low permeability porous cathode/anode structures are represented by the conventional Darcy equation. A computational sub-model describing the incorporation of electrochemical reaction source terms in the momentum balance equations for free and porous flow regimes within the fuel cell has been devised. The developed network permeability model and model have been incorporated in a general purpose single phase flow module to simulate the water flooding levels and dynamic flow field variables whilst the fuel cell is in operation. The network model is capable of simulating varying permeability in the porous regimes based on water saturation fluctuations. At increasing saturation, capillaries are blocked off progressively and the updated flow field is solved. Pore size distribution is a critical design parameter and henceforth a parametric study has been carried out to determine the effect of different pore-size distributions on the flow-field and cell efficiency in order to determine the optimum pore size structure. The simulated results for water saturation, dynamic flow field variables and fuel cell performance have been compared against the experimental data available in literature for purposes of model validation. The developed computer model offers a robust, cost-effective and user friendly design and analysis tool for fuel cell manufacturers, which can be easily used by engineers in industry.

1.0 Introduction

Proton Exchange Membrane Fuel Cells (PEMFCs) have the potential to become widely used portable power sources as they have a high power density, a high efficiency and create no on-site pollutants, making them ideal for numerous applications, such as in the automotive industry, or for powering mobile devices. They are also envisioned to be the next generation environmental friendly energy alternatives (Hirschenhofer, 1994; Fontes & Nilsson, 2001) as they have promising applications in the automotive and power sector due to factors such as low operating temperatures with zero/low emissions, a modular structure, quick start-up times, high power density and corrosion resistance

(Larminie & Dicks, 2000). Hence, efforts focusing on gaining an in-depth understanding of the dominant multi-physics within a fuel cell can aid in making them commercially viable and providing the desired power output/energy gains. Computational fuel cell models prove to be a valuable tool for engineers as they help in undertaking reliable simulations, and design optimization studies, facilitate the introduction of cheaper materials/fabrication techniques and provide cost/time effective solutions to appraise the design of complicated cell configurations. However, to date there have been no complete and generic computational PEM fuel cell models (Biyikoglu, 2005) that can account for all the complexities associated with the underlying multi-physics in a single mathematical framework.

In PEM fuel cells oxygen catalytically reacts with hydrogen, directly producing electrical power, heat and water. This water is either produced in liquid or vapour form on the cathode catalyst layer depending on the operating conditions, and the water vapour can condense within the cell. When too much liquid water forms, it causes a restriction in the flow of reactant gases through the porous zone microstructure to the catalyst layers, and a restriction in the exhaustion of water vapour into the gas channels. This is because the pathways through which the gas transport occurs become more tortuous (the opposite is true for liquid water). This phenomenon is known as flooding and causes serious performance limitations. Water management is thus very much crucial for the successful operation of the PEM fuel cell. To address the water management issues, this work considers pore scale models of porous materials, specifically PEM fuel cell Gas Diffusion Layers (GDLs), and the methods by which performance limiting parameters can be related to these model, namely network modelling and percolation theory.

Despite this, the majority of CFD models for PEM fuel cells either do not consider the effects of liquid-water flooding on cell performance or use simplistic assumptions. In the last decade or so there have been various analytical, semi-empirical, one-dimensional and two-dimensional PEMFC models that have been proposed by Bernardi *et al.*(1990, 1992), Springer *et al.*(1991), Fuller & Newman (1993), Nyugen *et al.*(1993), Amphlett *et al.*(1995), Standaert *et al.*(1996,1998), Bevers *et al.*(1997), Hubertus *et al.*(1998), Gurau *et al.*(1998), Yi *et al.*(1998), Eikerling *et al.*(1998), Wohr *et al.*(1998), Singh *et al.*(1999), Scott *et al.*(2000), Squadrito *et al.*(1999), Baschuk *et al.*(2000), Dannenberg *et al.*(2000), Hsing *et al.*(2000), Um *et al.*(2000), Rowe *et al.* (2001), Maggio *et al.*(2001), Wang *et al.*(2001), Eaton *et al.*(2001), Chan *et al.*(2003), Natarajan *et al.*(2003), Maxoulis *et al.*(2004) and Siegel *et al.*(2004).

Much literature exists on the general subjects of network modelling and percolation theory which can be referred to in the works of Fatt (1956), Bear *et al.* (1972, 1987), Arns *et al.* (2004), Suchamel *el al.* (1998), Reeves *et al.* (1996), Vogel *et al.* (2000), Garboczi *et al.* (1992), Selyacov *et al.* (1987, 1996), Glushco et al. (1989), Sahimi (1993), Strauffer & Aharony (1978), Tomadakis *et al.* (1993) and can be applied to model water flooding problem. Later some of the experimental work and porous media material modelling techniques were undertaken by Yeong & Torquato (1998). In context of PEMFC pore scale modelling approach for water management issues the only studies reported in the literature include those undertaken by Nam *et al.* (2003), Lister *et al.* (2006), Tseng *et al.* (2005), Passaogullari *et al.* (2004), Zhan *et al.* (2006), Mukherjee *et al.* (2006) and Baschuk *et al.* (2000).

In this work to model flooding within PEMFC's, it was decided to model relative, absolute permeability and effective diffusivities of species within the GDL material. This was done using a network modelling approach, based on the works of Fatt *et al.* (1956), Bear *et al.* (1972) and a percolation theory based approach following Selyacov et al. (1996). The presence of liquid water in the porous gas diffusion layer strongly affects the reactant gas, water vapour and liquid water permeability, and effective diffusivities. Hence models for permeability and effective diffusivity are an appropriate way of quantifying the effects of water saturation, and have been incorporated into a single phase CFD framework as presented in this paper. As the percolation theory and computational network modelling approaches both consider the formation of pathways in stochastic networks, the results using each of these methods should be consistent. As percolation theory based models tend to be used for infinite networks these models can be further used for optimization of the porous zone microstructures when coupled with the electrochemistry and the fluid dynamics in a PEMFC model.

2.0 Mathematical Model

The governing equations of the PEMFC model can be referred to in the recent work by Dawes, Hanspal & Turan (In Press 2009). The 3-D numerical model consists of two gas channels transporting hydrogen and oxygen reactants and water vapour (the product) mixed in air (nitrogen) to and from two porous gas diffusion layers (GDLs). The GDLs in turn transport the gases to and from the catalyst layers where all the electrochemical reactions occur. The catalyst layers sandwich a membrane through which protons and water are transported. The governing equations for the gas channels, GDLs and catalyst layers consist of continuity, conservation of momentum and species transport. Source terms are applied at the catalyst layers to represent the electrochemistry and transport through the membrane. In order to model these source terms the local current density (used to determine reaction rates) and net water transport coefficient (used to determine transport across the membrane) are calculated using auxiliary models which have been linked to the main flow solver through UDFs (User Defined Functions). The liquid phase of water is not directly modelled. As such, this simplification significantly reduces the complexity of the simulation; removing the need to model phase changes and allowing for an isothermal steady state simulation which is a valid assumption for low-temperature PEMFC applications. The flow hydrodynamics, electrochemistry and species transport are further coupled with pore scale and network modeling approaches for determining permeability and effective diffusivities as described in the works by Dawes, Turan & Hanspal (submitted, 2009).

3.0 Computational Results & Discussions

This section presents and compares results for the absolute and relative permeabilities generated using the percolation theory and network modelling methodologies described in Dawes, Turan & Hanspal (submitted, 2009). The percolation theory results are compared to experimental data Tseng *et al.* (2005). This process is not repeated for the network model as changes to the network topology require re-writing of the code rather than simply changing the percolation threshold (and critical exponent if the dimensionality of the network changes). Finally results for the percolation theory effective diffusivity model are presented and compared to experimental data, and other commonly used models.

3.1 Comparison of Percolation & Computational Results

The relative permeabilities generated by the network model are compared to the analytical expressions determined using percolation theory for a square network, shown in Fig.1, and a simple cubic network, shown in Fig.2. The parameters used for this parametric study are found in Table 1-4.

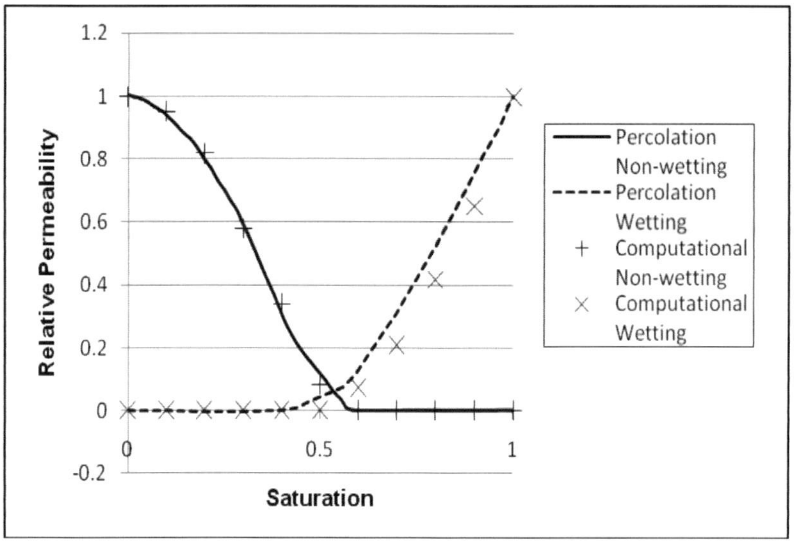

Figure 1: Regular square network

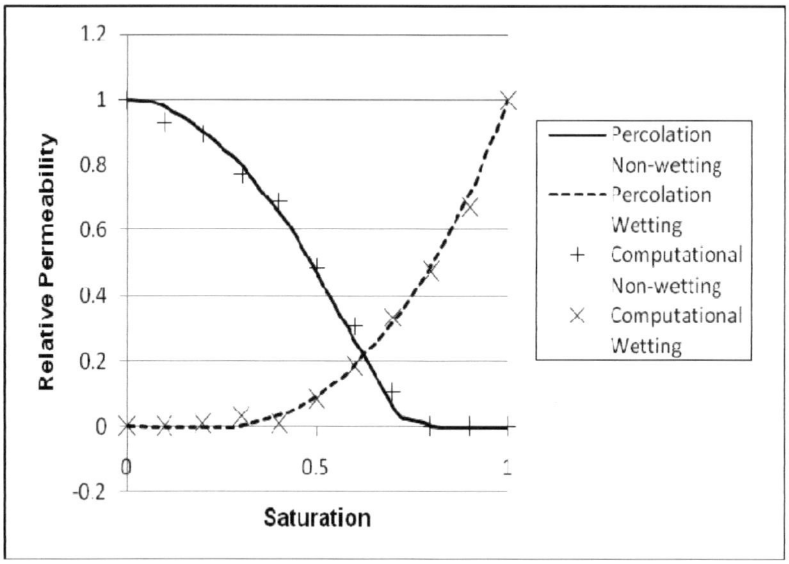

Figure 2: Simple cubic network

Table 1: Study Parameters

Mean bond diameter (μ)	21×10^{-6} m
Standard deviation (σ)	1.5×10^{-6} m
Porosity (ε)	0.75
Percolation threshold	0.5 (square), 0.25 (cubic), 0.11 (experiment)
Sample size (Computational)	8 x 8 x 8
Bond length (Simple cubic) (m)	3.722×10^{-5} m
Contact angle	110°

Table 2 Correlation index

2 dimensional	1.33
3 dimensional	0.90

Table 3 Bond Probabilities

Reactant gases	Liquid water
$P_b = \varepsilon$	$P_b = \varepsilon$
$P_b = 1 - S$	$P_b = S$

Table 4 Pore sizes

$f_1(x) = \dfrac{f(r)}{1 - \int_{r_k}^{\infty} f(r)dr}$	if $r > r_k$
$f_1(r) = 0$	if $r \leq r_k$

As the wetting fluid saturation increases, the non-wetting fluid relative permeability decreases from unity to zero. The network stops percolating the non-wetting fluid when the saturation radius reaches a critical value. In the regular square network case (Fig. 1), when the saturation increases further, the network begins to percolate the non-wetting fluid. For the simple cubic case (Fig. 2), there is an overlap range of saturations where both fluids are percolating through the medium. In both cases, the flow of each phase is heavily restricted by the presence of the other phase. In the two dimensional case, as can be seen in Fig. 1, the permeability of the medium is very strongly affected by saturation. This restriction affects the flow to such an extent that at most saturation values, only one phase is mobile in the medium.

For the simple cubic case, because more bonds are present in the medium, there are more pathways available to each fluid at each value of saturation, so the flow of each phase is less restricted. The results given by the computational method and those found using percolation theory match closely. The two methods considered the same network; however, the computational modelling solved a small stochastic network, whereas the percolation theory found analytical solutions for infinite networks. The permeabilities generated by the computational method were averaged from samples of 30 computations. As would be expected, larger networks produced the same mean relative permeabilities, but lower variance in the results. If the gas diffusion layer is treated with a hydrophobic coat, such as Teflon, liquid water will be the non-wetting phase. So the exhaust of liquid water, especially at low saturations, will be heavily affected by the percolation threshold, i.e. networks with higher coordination numbers will be less prone to flooding. For a pore structure with a low percolation threshold, liquid water will form a continuous fluid path through the medium at a lower saturation, and so liquid water will be exhausted more effectively.

3.2 Computational Model Results

A parametric study was undertaken via the computational model to determine relationships useful in GDL pore size distribution design. Relative permeabilities were found for four different normal probability density functions with low and high variance, and low and high mean. The parameters used are given below in Table 5.

Table 5: Variance Study Parameters

Case	Mean (μm)	Variance (μm)
Low Variance	20	1.5
High Variance	20	4.9
Low Mean	10	1.5
High Mean	30	1.5

Altering the mean pore radius of the sample only slightly altered the effects of saturation on the relative permeabilities. Increasing the mean slightly increased the non-wetting phase relative permeability and slightly decreased the wetting fluid relative permeability. However, the mean bond radius affected the absolute permeability to a large degree.

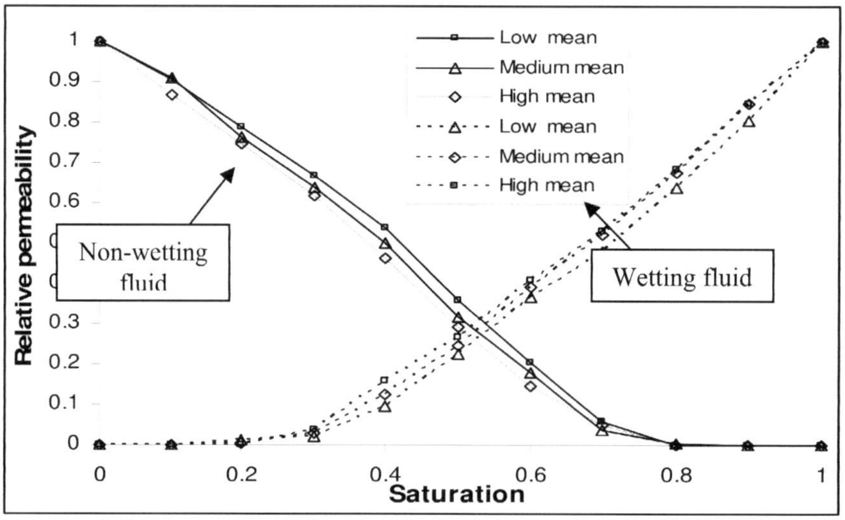

Figure 3: Effect of pore size mean

Increasing the variance was found to decrease the effects of wetting fluid saturation on the relative permeability of the non-wetting phase and reduce the values of relative permeability for the wetting phase permeability with respect to the wetting phase saturation. In hydrophobic gas diffusion layers liquid water is the non-wetting fluid, so increasing the variance aids the removal of liquid water. This is because with a larger variance there is a larger spread of pore sizes so more pathways through the network are available to the liquid water.

3.3 Percolation Theory Results

3.3.1 Comparison of Analytical Percolation Expressions to Experimental Data

Percolation theory was used to compare analytical results to experimental data. As mentioned previously, percolation theory was used because the only required bonding structure information was the percolation threshold, making the tailoring of the solution to a different network structure less time consuming than rewriting the computational code to take this into account. The percolation threshold for gas diffusion layer type material (randomly orientated fibres) was found by Tomadakis (1993) to be 0.11. To determine the bond length, the bonding structure was considered. The percolation threshold of the GDL is similar to that of a face centred bonding structure, so for the purposes of finding the average bond length in the gas diffusion layer, the structure was taken to be face centred cubic.

3.3.2 Absolute Permeability Results

Table 6 Percolation theory results

Model absolute permeability	$1.092 \times 10^{-12}\ m^2$
Experimental absolute permeability	$0.872 \times 10^{-12}\ m^2$

The actual randomly oriented fibre network has a lower percolation threshold, and so a higher coordination number. A higher coordination number will result in more capillaries occurring in the medium and so a larger total length of capillaries in a repeating cell. Thus with higher total capillary length, the bond length will be larger. This means that the assumption of similarity between the two networks will lead to an underestimation of the bond length. A larger bond length will result in a smaller permeability as reflected by the results in Table 6. This fact, along with the interflow of fluid between parallel capillaries, will contribute to the difference between the model and experimental absolute permeability.

3.3.3 Relative Permeability Results

Using the percolation threshold data from Tomadakis et al. (1993) analytical percolation results for the relative permeabilities in TGPH-120 carbon paper can be found (Figure 4).

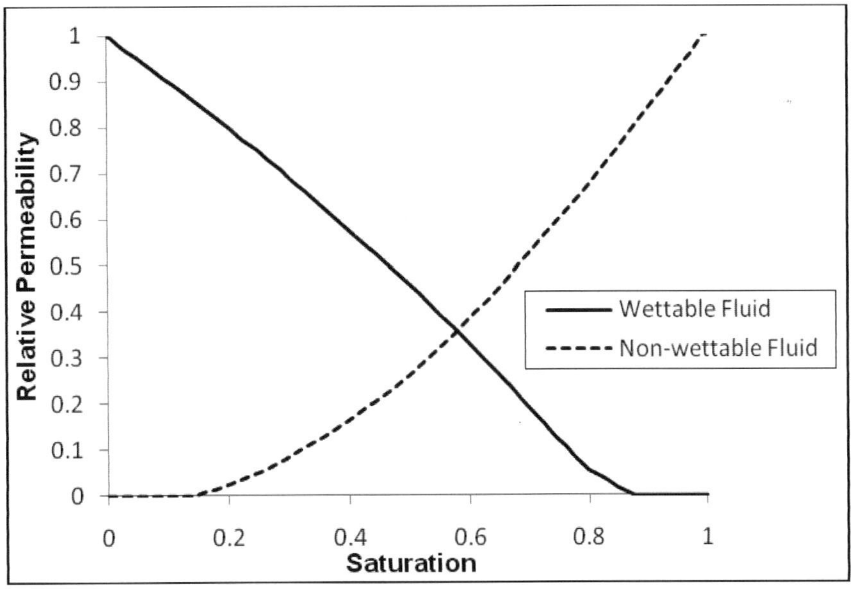

Figure 4: Relative permeabilities

These results show that the considered GDL material has a low percolation threshold, and so is conductive at low values of saturation. However, the effects of saturation occur quickly, i.e. there is no delay in the occurrence of the permeability drop due to saturation as seen in the simple cubic and regular square networks. It can also be seen that the liquid water relative permeability in this model will be zero until the saturation reaches approximately 0.11. This restriction on the removal of liquid water would lead to a build up of liquid water in the membrane. If transport due to diffusion was not considered for the liquid water, this limitation would mean that any liquid water produced would have built up in the GDL until the saturation reached 0.11 before a sample spanning cluster formed and percolation of liquid water out of the GDL could occur.

3.4 Diffusivity Results

A comparison was undertaken (Figure 5) between the analytically derived percolation expressions for effective diffusivity scaling functions (refer Dawes, Hanspal & Turan; 2009), computational results found by Nam *et al.* (2003) and the widely used Bruggeman correlation and a saturation power law approximation.

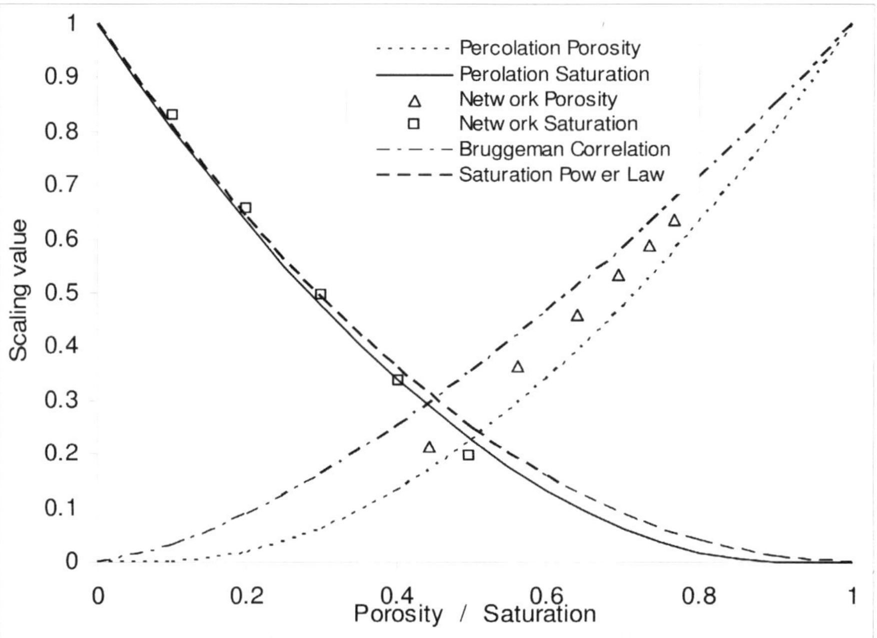

Figure 5: Comparison of diffusivity scaling methods

The percolation based relationships matched closely with the Bruggeman correlation (a percolation relationship for close packed spheres) for the effect of porosity and the $(1\text{-S})^2$ approximation for the effect of saturation. Closer agreement was found between the developed percolation theory based equations refer (Dawes, Turan & Hanspal; 2009) and the results generated by the diffusivity network model developed by Nam *et al.* (2003). This is because the network model and the developed percolation expressions considered similar microstructures (randomly oriented fibres and randomly shifted overlapping screens). The expressions show that a higher porosity will result in less of a reduction in the effective diffusivity. Effective diffusivity values are dominant in the flow of the reactant gasses in the gas diffusion layer. High porosity values and so high mean pore sizes will result in high diffusivities and so fast transport of oxygen to the catalyst layer and water vapour away from the catalyst layer. This will improve the performance of the fuel cell.

3.4 Numerical Simulations

In this section results are presented for water distributions and average current densities for a three dimensional CFD fuel cell model incorporating the percolation based diffusivity model derived in section 2.3. The developed model for species diffusivity was implemented in a 3-D single phase fuel cell model, using a 14114 hexahedral cell mesh, consisting of GDL membrane and gas channels. The fuel cell's electrochemistry was coupled to the governing equations by solving auxiliary equations defining the electrochemistry and membrane performance, thus calculating the local current density (used to determine reaction rates) and net water transport coefficient (used to determine transport across the membrane). Source terms were applied at the catalyst layers to represent the electrochemistry and transport through the membrane based on the calculated parameters. The model is similar to those developed by Springer *et al.* (1991), Dutta *et al.* (2000) and Lum *et al.* (2005) amongst others. The liquid water saturations for use in the water-flooding model are predicted using the relative-humidity values and water-vapour volume fractions. For further details of the numerical model please refer to Dawes, Hanspal & Turan. (2009).

3.41 Simulation Results

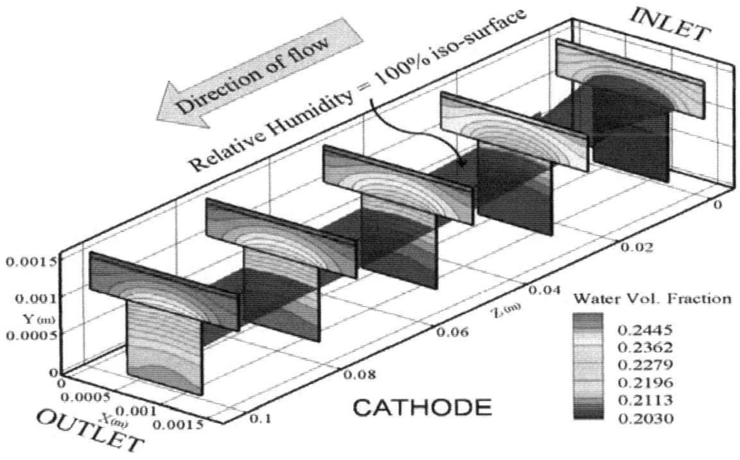

Figure 6: Water Volume Fraction at Cathode

Figure 6 shows the water volume fraction in the cathode, an isosurface is plotted for a relative humidity value of 100% with liquid water being above the isosurface. The case presented in this figure corresponds to the parameters (inlet mixtures species proportions and velocities) used in the very high humidity case of Lum *et. al* (2005), for which a marked deviation from experimental results was observed due to liquid water flooding. The figure shows that there is a relatively large amount of liquid water in the cathode GDL, with the saturation (or volume fraction of liquid water) being approximately 0.24, blocking the flow of reactant and exhaust gasses to and from the gas channel. This flow restriction accounts for the performance drop found experimentally.

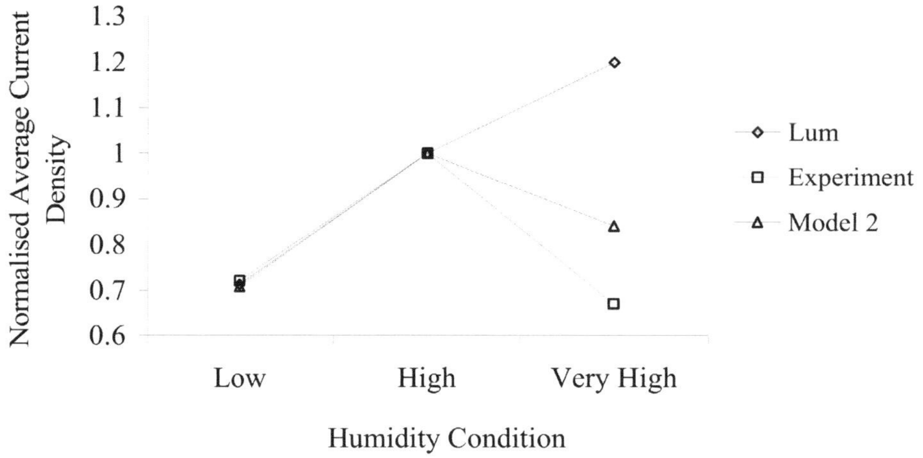

Figure 7: Global validation with variable inlet humidity, comparison against the numerical model and experiment data from Lum *et al.* (2005)

Results from our model have been validated against the experimental and computational data of and Lum *et al.* (2005) as shown in Figure 7. The results show a marked improvement in current density prediction at high humidity, where the liquid water flooding discussed above occurs.

4.0 Conclusions

Two different methods namely, a computational network model approach, and then an analytical approach using percolation theory was used in this work. Using these methods, the effect of the network structure on permeability and diffusivity was studied. Networks with a higher bond density, or a higher number of interconnections between pores, have a lower percolation threshold. This results in percolation occurring at lower saturations. So continuous percolating pathways of liquid water form between the catalyst layer and gas channels at lower saturations, and the membranes ability to exhaust liquid water is improved.

The pore size distribution can be optimised for liquid water transport. A parametric study was undertaken into the effect of mean and variance in the pore size distribution on the permeability and diffusivity. Increasing the pore size variance was found to improve the exhaustion of water in gas diffusion layers treated with a hydrophobic coating, as it allowed more pathways to form through the medium at all values of saturation. Increasing the mean pore size was found to increase the absolute permeability as each bond in a conducting chain had a higher value of conductance, however, it was not found to have a strong affect on the relative permeabilities.

Effective diffusivity is a critical parameter in fuel cell modelling as the flow of reactant gases in the porous zone is diffusion dominated. Diffusivity is strongly dependant on porosity and saturation. In the fuel cell, it was found that effective diffusivity can be increased by increasing the mean pore size (and so increasing the porosity), and also by reducing the liquid water saturation. Close agreement was found between the effective diffusivity expression derived using percolation theory and experimental data.

To gain more accurate insight as to the smaller scale flow features various improvements could be made, such as the use of more topologically accurate network models, generated from SEM (Scanning Electron Microscopy) scans or by using mean experimental coordination numbers or coordination number distributions. A simple improvement could be achieved by using the percolation threshold to estimate the mean coordination number for use in the network. Simple approximate relationships show that for a percolation threshold of 0.11 the mean coordination number will be approximately 13. Also, if the current network model were extended to be dynamic, considering unsteady percolation type water transport and incorporating condensation, large scale flow paths through the medium could be visualised and the nature of the displacement methods be better modelled. The dynamic model would not consist solely of drainage or imbibition, but would depend on the PEM fuel cell operating conditions.

The percolation theory equations (listed in Dawes, Turan & Hanspal; 2009) were directly incorporated into a 3-D PEMFC model and a marked improvement was found when accounting for liquid water flooding in the cathode using the diffusion model alone. Further use of these models and the extension of the CFD model to multiphase scenarios would allow optimization of material parameters which could be used in the development of flood resistant gas diffusion layers.

5.0 References

Amphlett, J.C., E.H. de Oliveira (1997). Dynamic interaction of a proton exchange membrane fuel cell and a lead-acid battery. Journal of Power Sources 65(1-2); 173-178.

Arns, J, Robbins, V, Sheppard, A, Sok, R, Pinczewski, W, Knackstedt, M. (2004) Effect of Network Topology on Relative Permeability, Transport in Porous Media 55; 21-46

Baschuk, J.J., X., Li (2000). Modelling of polymer electrolyte membrane fuel cells with variable degrees of water flooding. Journal of Power Sources 86(1-2): 181-196

Baschuk, J, Xinguo, L, (2000) Modelling of polymer electrolyte membrane fuel cells with variable degrees of water flooding. Journal of power sources 86 181-196

Bear, J. (1972). Dynamics of fluids in porous media, American Elsevier Publishing Company, Inc.

Bear, J, Braester, C, Menier, C. (1987) Effective and Relative Permeabilities of Anisotropic Porous Media, Transport in Porous Media 2, 301-316

Bernardi, D.M. (1990). Water-Balance Calculations for Solid-Polymer-Electrolyte Fuel Cells, J. Electrochem. Soc. 137(11): 3344-3350

Bernardi, D., M., M.W. Verbrugge (1992). A Mathematical Model of the Solid Polymer Electrolyte Fuel Cell, J. Electrochem. Soc. 139(9): 2477-2491.

Bevers, D., M. Wohr, (1997). Simulation of a polymer electrolyte fuel cell electrode. Journal of Applied Electrochemistry 27(11): 1254-1264

Bıyıkoğlu A. (2005). Review of proton exchange membrane fuel cell models. International Journal of Hydrogen Energy 30(11): 1181-1212

Chan, S.H., S.K. Goh, S.P. Jiang, (2003). A mathematical model of polymer electrolyte fuel cell with anode CO kinetics. Electrochimica Acta 48(13): 1905-1919

Dannenberg, K., P. Ekdunge, (2000). Mathematical model of the PEMFC. Journal of Applied Electrochemistry 30(12): 1377-1387.

Dawes, J; Hanspal, N; Turan, A; Family, O. 3e-Dimensional CFD Modelling of PEM Fuel Cells: An Investigation into the Effects of Water Flooding, Chemical Engineering Science (accepted, 2009)

Eaton B (2001). 1-Dimensional, transient model of heat, mass, and charge transfer in a proton

exchange membrane, Advanced Energy Systems Division, American Society of Mechanical Engineers; 41: 429–440.
Eikerling, M.; A. A. Kornyshev (1998). Modelling the performance of the cathode catalyst layer of polymer electrolyte fuel cells, Journal of Electroanalytical Chemistry 453(1-2): 89-106.
Fatt, I. 1956a. The network model of porous media. I. Capillary pressure characteristics, Petroleum Transactions, The American Institute of Mining, Metallurgical, and Petroleum Engineers
Fatt, I. 1956b. The Network Model of Porous media. II. Dynamic properties of a single size tube network, Petroleum Transactions, The American Institute of Mining, Metallurgical, and Petroleum Engineers
Fatt, I. 1956c. The network model of porous media. III. Dynamic properties of networks with tube radius distribution, Petroleum Transactions, The American Institute of Mining, Metallurgical, and Petroleum Engineers
Fontes & Nilsson (2001), Industrial Physicist, tipmagazine.com
Fuller, T.F.; J. Newman (1993). Water and Thermal Management in Solid Polymer Electrolyte Fuel Cells, J. Electrochem. Soc. 140(5): 1218-1225
Garboczi, E; Bentz, D. (1992) Computer simulation of the diffusivity of cement-based materials. Journal of Material Science 27, 2083 – 2092
Glushko, V; Selyacov, V; Kadet, V. (1989) Percolation model of two-phase equilibrium flow in a porous medium with microheterogeneous wettability, Fluid Dynamics, 24:5, 727-734.
Hirschenhofer, J.H., D.B. Stauffer, (1994). Fuel cells: A handbook (Revision 3)
Hsing, I.M.; P. Futerko (2000). Two-dimensional simulation of water transport in polymer electrolyte fuel cells. Chemical Engineering Science 55(19): 4209-4218.
Kadet V; Selyacov, V. (1987) Percolation model of two-phase flow through porous media, Fluid Dynamics, 22:1, 72-82.
Kadet, V; Popov, A; Selyakov, V. (1991). Effect of plastic properties on phase permeabilities, Fluid Dynamics, 26:2, 247-251.
Lister, S; Sinton; Djilali, N (2006). In-situ visualization of liquid water transport in PEM fuel cell gas diffusion layers, Journal of Power Sources, 154, 95-105.
Lum K.,W; J.J McGuirk, (2005) Three-dimensional model of a complete polymer electrolyte membrane fuel cell - model formulation, validation and parametric studies, Journal of Power Sources 143(1-2): 103-124.
Maggio, G.,V. Recupero, (2001). Modeling polymer electrolyte fuel cells: an innovative approach, Journal of Power Sources 101(2): 275-286.
Maxoulis, C.N., D.N. Tsinoglou (2004). Modeling of automotive fuel cell operation in driving cycles, Energy Conversion and Management 45(4): 559-573.
Mukherjee, P; Wang, C. (2006) Stochastic Microstructure Reconstruction and Direct Numerical Simulation of the PEFC Catalyst Layer, Journal of Electrochemical Society (2006), 153:5, A840-A849
Nam, J; Kaviany, M. (2003) Effective diffusivity and water-saturation distribution in single and two-layer PEMFC diffusion medium, International Journal of Heat and Mass Transfer 46, 4595–4611
Natarajan, D., T. Van Nguyen (2003). Three-dimensional effects of liquid water flooding in the cathode of a PEM fuel cell. Journal of Power Sources 115(1): 66-80
Nguyen PT, Berning T, Djilali N. (2004) Computational model of a PEMFC with serpentine gas flow channels. J. Power Sources 130: 149–157
Nguyen T.V., White R.E. (1993). A water and heat management models for proton-exchange-membrane fuel cells, J. Electrochem. Soc. 140: 2178–2186
Pasaogullari, U; Wang, C. (2004). Liquid Water Transport in Gas Diffusion Layers of Polymer

Electrolyte Fuel Cells. Journal of the Electrochemical Society, 151 (3) A399-A406
Reeves, P; Celia, M. (1996). A functional relationship between capillary pressure, saturation, and interfacial area as revealed by a pore-scale network model, Water Resources Research, 32: 2345-2358.
Rowe, A.; X. Li (2001). Mathematical modeling of proton exchange membrane fuel cells. Journal of Power Sources 102(1-2): 82-96
Sahimi, M; (1993) Flow phenomena in rocks: from continuum models to fractals, percolation, cellular automata, and simulated annealing. Rev. Mod. Phys. 65 (1993), 1393-1534
Scott, K. (2000), Modeling transport phenomena and performance of Direct Methanol Fuel Cell Stacks, Chem. Engg. Res & Des, 78: 881-888
Selyakov, V; Kadet, V. (1996) Percolation Models for Transport in Porous Media with Applications to Reservoir Engineering, Kluwer Academic Publishers
Siegel, N.P., M.W. Ellis, (2004). A two-dimensional computational model of a PEMFC with liquid water transport. Journal of Power Sources 128(2): 173-184
Springer, T.E., T.A. Zawodzinski, S. Gottesfeld (1991) Polymer Electrolyte Fuel Cell Model. Journal Electrochem. Soc. 138(8): 2334-2342
Squadrito, G., G. Maggio, (1999). An empirical equation for polymer electrolyte fuel cell (PEFC) behaviour. Journal of Applied Electrochemistry 29(12): 1449-1455
Standaert, F., K. Hemmes, (1996). Analytical fuel cell modeling. Journal of Power Sources 63(2): 221-234
Standaert, F., K. Hemmes, (1998). Analytical fuel cell modeling; non-isothermal fuel cells. Journal of Power Sources 70(2): 181-199
Strauffer, D; Aharony, A. (1978) Introduction to Percolation Theory', Taylor and Francis Ltd.
Suchmel, B; Chen, B; Allen, M. (1998) Network Model of Flow, Transport and Biofilm Effects in Porous Media, Transport in Porous Media (1998) 30: 1-23
Tomadakis, M; Sotirchos, S. (1993). Ordinary and Transition Regime Diffusion in Random Fibre Structures' AIChE Journal March, 39(3)
Tseng, C; Lou, S; Yan, Y; Sung, L. (2005). Gas Diffusion Layer and Microporous Layer for PEMFC Proceedings International Hydrogen Energy Congress and Exhibition IHEC Istanbul, Turkey
Um S, Wang CY, Chen KS. (2000) Computational fluid mechanics of proton exchange membrane fuel cells. J Electrochem Soc 147:4485–4493
Um S., C. Y. Wang (2004). Three-dimensional analysis of transport and electrochemical reactions in polymer electrolyte fuel cells. J Power Sources 125(1): 40–51
Vogel, H; (2000). A numerical experiment on pore size, pore connectivity, water retention, permeability, and solute transport using network models, European Journal of Soil Science, 51: 99-105
Wöhr M., K. Bolwin, W. Schnurnberger, M. Fischer, W. Neubrand and G. Eigenberger (1998) Dynamic modelling and simulation of a polymer membrane fuel cell including mass transport limitation" International Journal of Hydrogen Energy, 23(3): 213-218
Yeong C; Torquato, S. (1998). Reconstructing random media, Physical Review E, 57:1, 495-506.
Yi J.S., T.V. Nguyen (1998). An along the channel model for proton exchange membrane fuel cells. J Electrochem Soc 145(4): 1149–1159.
Yi, S., J.D. Yang, C. King, (2004). Water management along the flow channels of PEM fuel cells, AIChE Journal 50(10): 2594 - 2603
Zhan, Z; Xiao, J, Dayong, L; Pan, M; Y Rnuzhang (2006) Effects of porosity distribution variation on the liquid water flux through gas diffusion degree of PEM fuel cells, Journal of Power Sources, 160:2, 1041-1048

55a The U.S. Dept of Energy Hydrogen Program: Status and Performance Gaps of On-board Hydrogen Storage Technologies

Grace Ordaz, Ned Stetson, and Carole Read. Office of Hydrogen, Fuel Cells and Infrastructure Technologies, U.S. Department of Energy, 1000 Independence Ave., SW, Washington, DC 20585

The U.S. DOE Hydrogen Program's mission is to reduce oil use and carbon emissions in the U.S. transportation sector and to enable clean, reliable energy for stationary and portable power generation. The requirements for on-board vehicular hydrogen storage continue to be one of the most technically challenging barriers to the widespread commercialization of hydrogen fueled vehicles. The DOE-EERE hydrogen storage activity primarily focuses on applied research and development of materials and engineering to provide a low-pressure on-board system that meets packaging, cost, safety and performance requirements competitive with current vehicles in the North American market while providing at least a 300 mile (500 km) driving range.

This presentation summarizes the status, recent accomplishments and current performance gaps of hydrogen storage technologies for transportation applications. The research is conducted under framework of DOE's National Hydrogen Storage Project that includes independent projects and Centers of Excellence in both applied and basic hydrogen storage R&D. Material projects are focused in three main areas: metal hydrides; chemical hydrogen storage materials and hydrogen sorbents.

There are several new efforts, including the Hydrogen Storage Engineering Center of Excellence, which will provide a coordinated approach to engineering research and development of on-board storage systems.

Conversion of High Sulfur Fuels to Synthesis Gas for Fuel Cells

Lyman J. Frost, S. Elangovan, Joseph Hartvigsen; Ceramatec, Inc; Salt Lake City, Utah, USA.

Abstract

The conversion of high sulfur military fuels to synthesis gas (i.e. hydrogen and carbon monoxide) that can be used by fuel cells of any type is a difficult process. While solid oxide fuel cells (SOFC) are capable of using both hydrogen and carbon monoxide, the cells are generally still sensitive to the sulfur in the fuel. In addition, the military fuels are generally a mix of heavier hydrocarbons that can generate carbon if the reactions are not in the proper regime.

This paper will describe a non-thermal plasma reformer that generates a stream of synthesis gas appropriate for a SOFC auxiliary power unit from and input of military grade diesel fuel (e.g. JP8). Military fuels such as JP8 can contain up to 3000 parts per million sulfur. Typical catalytic based reformers are also sensitive to sulfur requiring the removal of sulfur by hydro-desulfurization or a liquid phase sulfur removal step. The non-thermal plasma approach has the advantage that it is not sensitive to sulfur. The non-thermal plasma acts in much the same way as a conventional catalyst to promote the reforming of the fuel but does not deactivate over time. It is in effect a constantly renewing catalyst.

Under proper conditions the non-thermal plasma can also be used to disassociate hydrogen sulfide into hydrogen and elemental sulfur. Using a cascade effect it is possible to reduce the amount of sulfur transmitted to the SOFC portion of the system to an amount that can be tolerated by the SOFC. This process will also be described in the paper.

Introduction

Fuel cells are well known to have significant potential benefits such as high fuel efficiency, low emissions, long life, and near silent operation. For the military, the fuel that can be used by the fuel cells will be a key determinant in how quickly fuel cells can be introduced into operational status. Current logistic fuels (e.g. JP-5, JP-8, JP-10, etc) are the desired fuels for military fuel cell systems. The ability to convert these fuels into a synthesis gas that can utilized directly by fuel cell systems is a critical technology. Ceramatec has developed an advanced technology that can convert the fuels currently available into the synthesis gas required as the basic feed to fuel cell systems. Coupled with appropriate downstream processing, the Ceramatec non-thermal plasma reformer can service any type of fuel cell system.

The Ceramatec non-thermal plasma system uses a novel patented process called GlidArc®. The gliding plasma arc creates radicals within the vaporized fuel stream to promote breaking of chemical bonds. This unique approach can process a wide variety of fuels and provides for saturation of aromatics (i.e. hydrogenation), liberation of deeply bound sulfur (if present), and hydrolysis (hydro-cracking) of large hydrocarbons. The Ceramatec plasma reformer is fuel flexible, sulfur tolerant, and compact in size. The Ceramatec plasma technology is an innovative, integrated process that exploits the thermodynamic advantages of steam reforming and combines it with the robust, highly energetic reaction zone characteristics of partial oxidation.

Ceramatec has tested its plasma reformer on a variety of fuels including diesel, JP-8, JP-10 and bio-fuels. The reformer does require an electric input to generate the plasma but since the plasma is non-thermal in nature, this requirement is minimal. Data obtained in tests with the plasma reformer indicate that ~1 – 2 % of the energy content of the fuel is required to provide electricity to the plasma reformation process (assuming a 50% efficient fuel cell). The radicals generated by the plasma act in much the same way as a solid catalyst by reducing the energy of activation and thus increasing the chemical reaction rate under Arrhenius's law. A further benefit of the Ceramatec plasma reformer is the variable operating temperature which is set by equilibrium thermodynamics. For heavy liquid hydrocarbon fuels the operating range is typically between 700-900°C, matching the operating range of high temperature fuel cells (e.g. SOFC) and aligning well for thermal integration.

Technology description

The objective of the reformer is to convert the chemical energy contained in various liquid fuels (e.g. diesel, JP-5, JP-8) into a synthesis gas feed specific to use in fuel cells. The gas temperature of non-thermal plasmas is very convenient for ignition and combustion support, as well as for fuel conversion. The generated radicals have a relatively long life time and cause a catalytic effect. These plasmas are powerful enough to provide fast start and short transition times between operating states. Non-thermal plasmas can support two different processes of fuel conversion: partial oxidation that is slightly exothermic and steam fuel conversion that is endothermic.

The conventional "flat" gliding arc (Fig. 1) starts as an electrical breakdown in a narrow gap between two or more diverging electrodes in a gas flow. When the electric field in this gap reaches the critical level, the current of the arc increases very fast and the voltage of the arc drops. If the gas flow is strong enough, it forces the arc to move along the diverging electrodes and to elongate. At some point the resistance from the divergence of the electrodes extinguishes the arc and a new arc forms at the narrow gap. Also, in contrast to the thermal arcs cooled predominantly by conductive heat transfer, non-equilibrium discharge is "ventilated" by dragging flow (convective cooling), and becomes wider and less bright (Fig. 2).

Figure 1: Gliding arc discharge in air

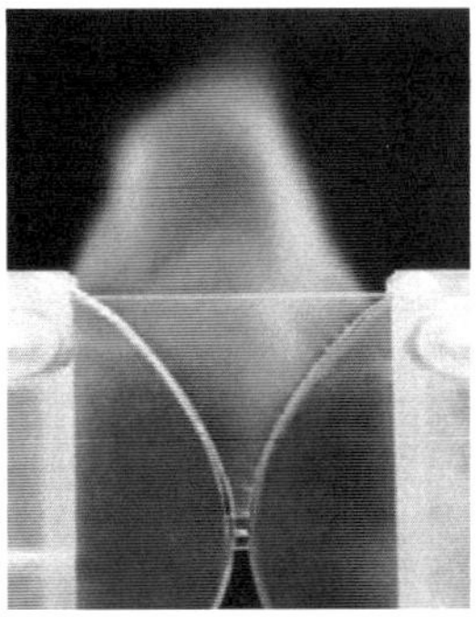

Figure 2: Gliding arc evolution at 5 ms separation between pictures

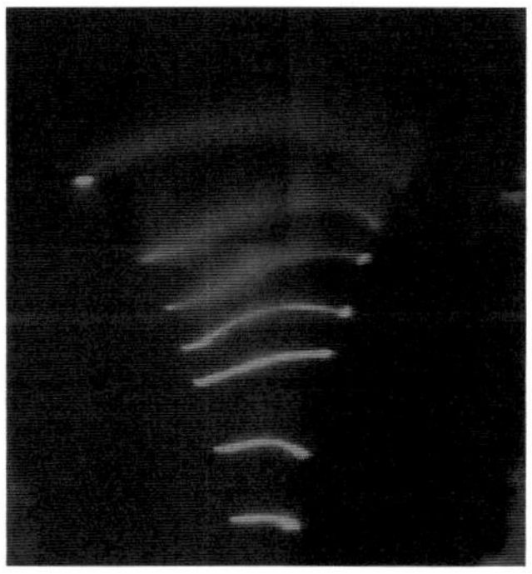

Ceramatec has demonstrated the GlidArc® reformer in both the exothermic partial oxidation mode and endothermic steam reforming mode to provide a flexible reformer. The reformer can start rapidly in a partial oxidation mode and as system heat becomes available switch to the more efficient steam reforming mode.

Ceramatec has tested its low-power, non-thermal, *plasma-catalyzed* reformer with widely available logistic fuels, such as diesel and JP-8. The plasma reforming technology is an innovative, integrated process that retains the thermodynamic advantages of steam reforming and combines it with the robust, highly energetic reaction zone characteristics of partial oxidation. The process is easily scalable for use in compact power systems, ranging from a few kilowatts to hundreds of kilowatts. The laboratory-scale reformer (Figure 3) is compact, with an internal volume of 0.6 liter and a reformation capacity of 5 kWt fuel rate.

The plasma reformer generates a highly active, non-thermal plasma source of non-equilibrium electric discharges that act as an activation source for plasma chemical reactions. The operating temperature is set by the thermodynamics of the reforming reaction, which typically means that temperatures in excess of 800°C are required for high conversion of hydrocarbons to synthesis gas. Reforming reactions are highly endothermic, consuming up to 30% of the heat of combustion of the fuel. This reformation heat is not supplied by the low-energy plasma, but may be supplied by either partial oxidation of the fuel within the reformer or process integration with other temperature matched heat sources. Fortuitously, the reforming temperature corresponds with the heat rejection from high temperature fuel cells, allowing opportunities for thermal integration.

The device has at least two diverging knife-shaped electrodes along which multiple electric discharges glide. These electrodes are immersed in a fast-flowing gas. The progress of an electric discharge is shown in Figure 3 (right). A high-voltage, single- or poly-phase, relatively low-current discharge is generated across a gas flow between the electrodes. The discharge forms at the closest point, spreads along the electrodes, disappears, and another discharge immediately forms at the initial spot. The geometry of the electrodes, flow conditions, and characteristics of the power supply determine the path of the arc. The electrodes do not need to be cooled, so the electrical energy is directly and totally transferred to the processed gas. The voltage can be as high as 20,000 volts. Multiple discharge systems can be installed for large gas flow rates. Any gas or vapor can be directly processed. Droplets, mist, and powders can be present without the risk of soot formation.

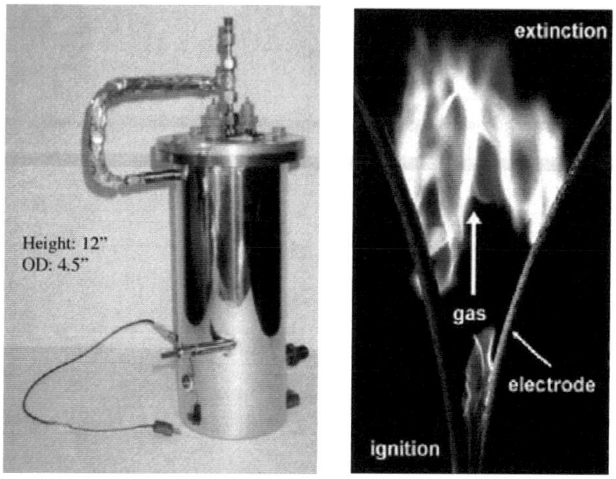

Figure 3: GlidArc ® plasma reformer: compact 5-kWt JP-8 reformer (left); inside view of the GlidArc ® (right).

JP-8 fuel obtained from a Chevron refinery was reformed using the reformer in a partial oxidation mode. The analysis provided by the refinery indicated that the fuel had a total sulfur of 430 ppmw and an aromatic content of 15.3 vol%. A reformate gas analyses using air for the oxygen supply performed over a 16-hour operation is shown in Figure 4. The H_2 + CO content of the reformate was around 30 vol%. Even with the 15% aromatic content, there was only ~1.5 vol% of methane in the reformate and negligible higher hydrocarbons.

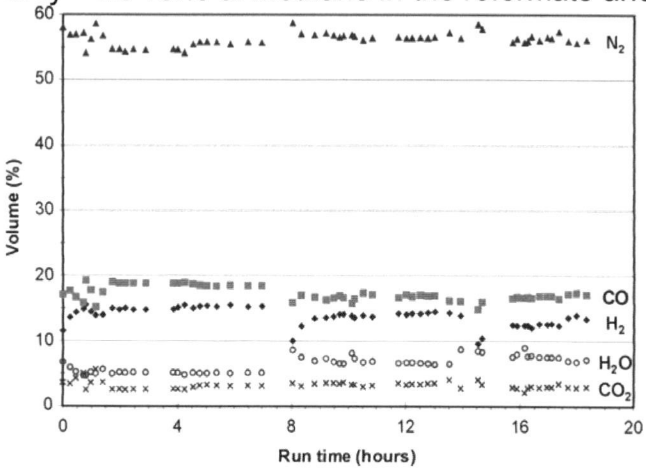

Figure 4: Gas composition of JP-8 reformate.

Figure 4 shows the percentages of hydrogen and carbon monoxide generated. The constituents of the generated synthesis gas behave as expected, with hydrogen rising as water content drops and carbon monoxide percentages dropping when carbon dioxide percentages increase. The carbon dioxide and water occur in all reactions of this type since the reaction kinetics are not 100% selective for the desired hydrogen and carbon monoxide. Once created, the hydrogen and carbon monoxide also will react with any available oxygen to form the water and carbon dioxide. The percentages of gas content are consistent with the use of air as the oxidizing medium and JP-8 as the reformed fuel. There is of course some moisture and carbon dioxide in the input air and this is reflected in the final gas percentages.

The plasma reformers are compact, easily scalable, and the electrical consumption is low (i.e. estimated to be in the 1–2% range of the fuel cell's full load output) enabling high overall system efficiency. Tests have successfully reformed pipeline-grade natural gas, commercial propane, gasoline, and diesel fuel at atmospheric pressure. The electrical power was ~90 W for a 5-kWe fuel rate (about ~2% of the fuel cell output).

Plasma Catalyzed Steam Reformer Design and Performance

The partial oxidation mode GlidArc® reformer is designed to minimize heat loss from the reaction zone, and to recover some of the heat loss in preheating the reactant air stream. Operating in an endothermic mode, the reformer must provide the required heat transfer duty of 25-30% of the feed stream heat of combustion for the reforming reaction. It is desirable to preheat the steam/air oxidant mixture to near reforming temperature since heat transfer area in the plasma zone is very limited. A large post plasma zone heat transfer area is also provided to allow further conversion.

Results of JP-8 testing using the endothermic plasma catalyzed steam reformer design show the improved performance with respect to reformate composition (Figure 5) and reformate heating value efficiency (Figure 6) relative to operation in partial oxidation mode.

Figure 5: Reformate composition; partial oxidation and steam reforming modes

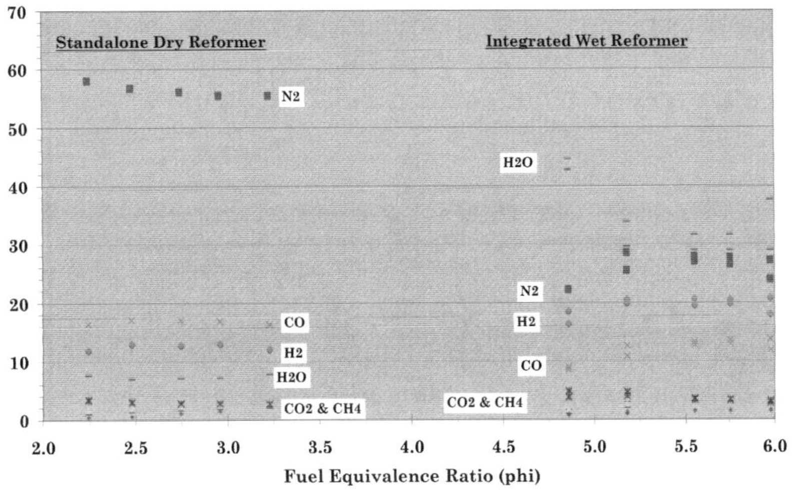

Figure 6: Efficiency of partial oxidation and steam reforming modes

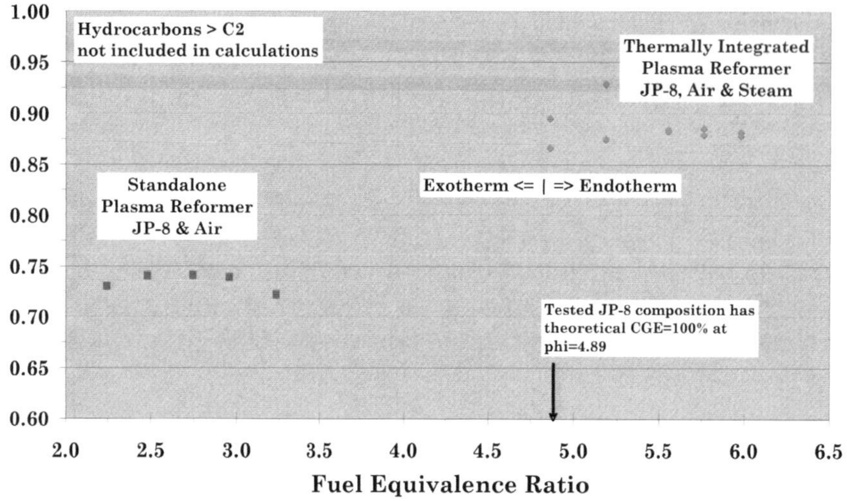

Ceramatec's compact advanced non-thermal plasma reformer requires no cooling, uses simple power supplies, experiences insignificant electrode corrosion, and can be easily integrated into other processes. The system may be modularized and scaled up using multiple-electrode configurations. The powerful, inexpensively produced, and easily controllable gliding discharges produced in the plasma reformers can be used as a low-cost method of producing synthesis gas from hydrocarbons and an oxidant (e.g. steam or air). The gliding discharge system has no thermal inertia and can therefore respond immediately to control signals. No catalysts are required, which significantly improves its ability to handle liquid logistic fuels and makes pretreatment of the fuel to remove sulfur unnecessary.

Disassociation of hydrogen sulfide

It is also possible to use the non-thermal plasma unit to disassociate hydrogen sulfide into hydrogen and elemental sulfur. If operated at the proper temperatures, this approach has two substantial advantages: the production of additional hydrogen and the removal of sulfur as a solid product.

Traditional methods for removal of sulfur usually involve some variant of the Claus sulfur removal process. The Claus process involves separation of hydrogen sulfide, controlled burning of the hydrogen sulfide at high temperature, and then reaction of a mixture of hydrogen sulfide and sulfur dioxide over a series of catalyst beds. Even the best catalyst beds leave some hydrogen sulfide and compounds formed in side reactions that must be removed with a tail gas process such as the Shell Catalytic Offgas Treatment (SCOT) process. These processes are large, expensive to operate, and complex in nature.

The plasma sulfur removal unit follows a flow pattern as shown in Figure 7. This diagram shows the potential flow of the gasification stream to the plasma reformer and the process for recovering the sulfur. The unit has only been tested for limited periods of time on the recovery of sulfur from hydrogen sulfide but the data gathered to date is shown in Figure 8. It is felt that the amount of sulfur removed can be improved. The reaction is as follows:

$$H_2S + \text{electric} + \text{heat} \rightarrow H_2 + S$$

An advantage of this process is the ability to operate at temperature with little cooling of the synthesis gas. The commercial processes generally used (e.g. Claus) involve cooling and reheating of the gas and associated process streams. Even with process integration these steps reduce the overall efficiency of the system. The use of the plasma methodology for removal of hydrogen sulfide is expected to improve overall system efficiency.

Figure 7: Flow diagram of hydrogen sulfide to non-thermal plasma reformer

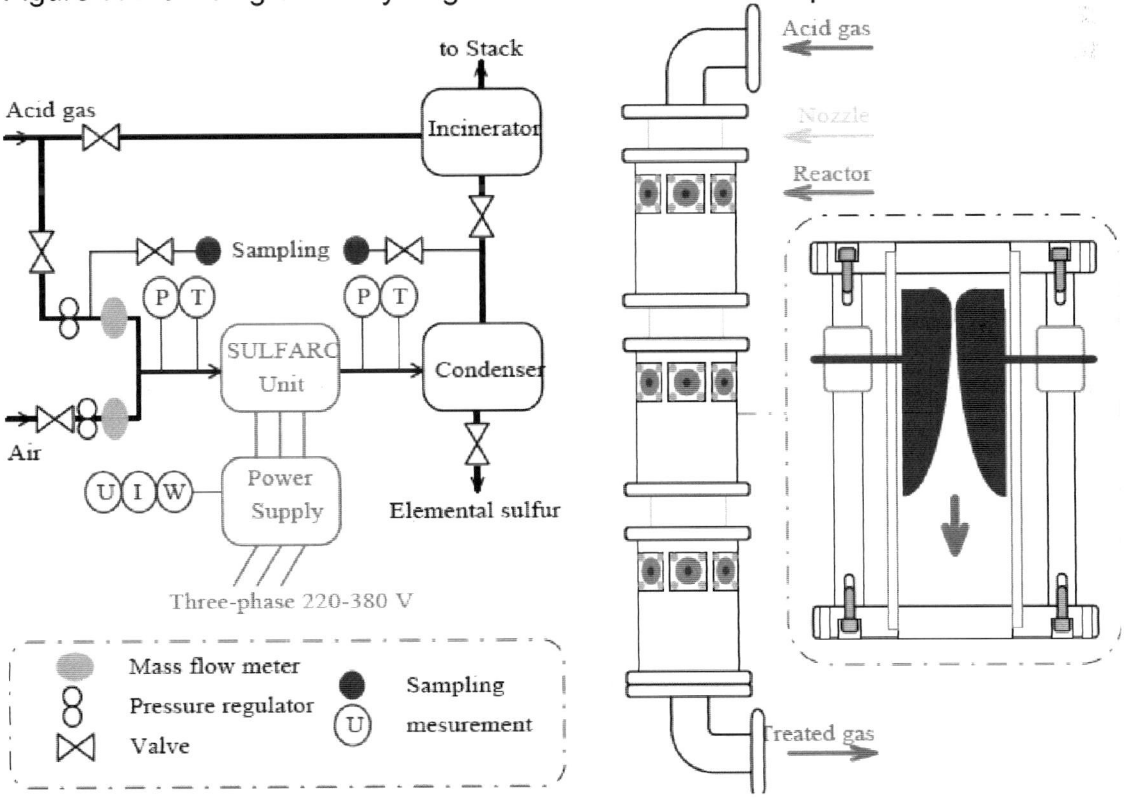

Figure 8: Results of non-thermal plasma reformer operation on hydrogen sulfide

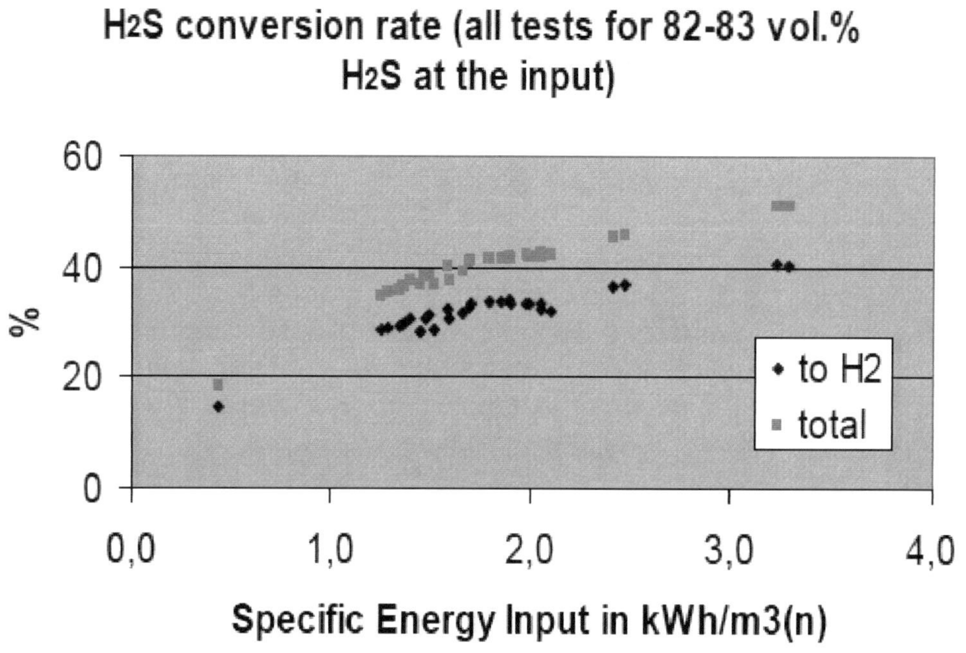

The work related to sulfur removal shown above has been done in consultation with Dr. Albin Czernichowski of France.

Hydrogen Production via Solar Water-splitting Using the Hybrid Sulfur Process

William A. Summers[1], *Claudio Corgnale*
Savannah River National Laboratory
Aiken, South Carolina, USA

One of the most promising approaches for addressing the growing concerns related to the world's current fossil fuel energy system is the transition to a sustainable system based on the use of electricity and hydrogen as the primary energy carriers. Hydrogen is plentiful, sustainable, energy efficient and extremely friendly to the environment. When hydrogen is consumed in a fuel cell engine, the reaction with oxygen gives only pure water as exhaust product. The widespread use of hydrogen as an energy carrier has often been described as the "hydrogen economy."

However, hydrogen is a very reactive element, and it does not occur in significant quantities on earth in a pure state. Most hydrogen is combined with other elements and is found in water, fossil fuels, biomass and similar compounds. A hydrogen economy requires primary energy sources to provide the energy to release the hydrogen from these compounds. Water-splitting processes are particularly attractive, since water is abundant around the planet, the unique products of the process are hydrogen and oxygen and a variety of primary energy sources can be used to produce the hydrogen. External energy can be used to produce hydrogen by conventional water electrolysis (1), which first requires the production of electricity. A more efficient and potentially more cost effective approach is to use external thermal energy to power a thermochemical process, which breaks the hydrogen-oxygen bounds through a series of heat-driven chemical reactions, with recirculation of all intermediate substances (2), (3), (4).

This paper will discuss research activities on the Hybrid Sulfur (HyS) thermochemical water splitting process carried out at Savannah River National Laboratory (SRNL). The HyS process is an advanced and promising cycle that has been extensively studied and experimentally verified. It is one of the simplest thermochemical processes, comprising only two global reaction steps and having only fluid reactants. A simple schematic of the cycle is shown in the **Figure 1**. The high temperature reaction step for the decomposition of concentrated H_2SO_4 (usually 70–90wt% in H_2O) into SO_2, O_2 and H_2O is common to all sulfur-based thermochemical cycles. It's an endothermic reaction which requires high temperatures (usually higher than 800 °C) to make the process feasible. After O_2 is removed from the process (as a co-product), SO_2 and H_2O are recycled and electrochemically combined again, feeding the anode of a SO_2 Depolarized Electrolyzer (SDE) to produce H_2 and H_2SO_4 which is then recycled to the high temperature decomposition step, closing the whole cycle.

[1] Corresponding author Tel: +1 803 7257766 Fax: +1 803 7258829
Email address: william.summers@srnl.doe.gov

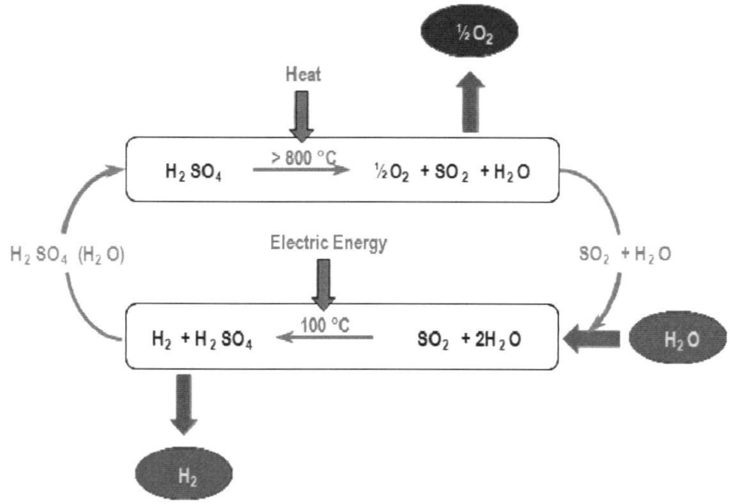

Figure 1: The HyS cycle

The HyS cycle was developed starting from the 1970's (5), (6) and extensive research and development has been restarted in the last few years (3), (7) to assess the effective potential to be coupled to both nuclear and solar source power. The SRNL has been developing the HyS cycle since 2004 as part of the DOE's Nuclear Hydrogen Initiative (NHI). SRNL developed and constructed a new SDE concept which has successfully been tested for a H_2 production rate of 100 l/h in a 3 cell stack arrangement (7), (8), (9). SRNL is also involved in the DOE Solar Thermochemical Hydrogen (STCH) project and recently completed the conceptual design and economic analysis for a HyS process matched with a solar central receiver plant and sized for a yearly average H_2 production of 100 tons/day.

To evaluate the performance and economic potential of the solar HyS plant in the medium to long term periods, two scenarios have been identified. The first one sees the 2015 as the construction starting year, while the second scenario, for the long period, the year 2025. The solar receiver design is based on a falling particle receiver developed by Sandia National Laboratory using sand as the intermediate heat transfer medium (10). The coupling scheme between the sand particle receiver solar tower plant and the HyS plant interfaced equipment (i.e. the high temperature H_2SO_4 decomposition bayonet reactor and the reboiler of the H_2SO_4 concentration column), shown in **Figure 2**, has been assumed as the baseline concept for the 2015 year. A sand flow rate goes down inside the solar tower, is heated up to 1000 °C by concentrated solar radiation in the particle receiver and is then stored in the hot sand storage tank. Sand passes from the hot tank across an intermediate heat exchanger, IHX (sand to fluid h.e. in **Figure 2**), to a cold storage tank. A secondary intermediate fluid (e.g. helium, air) is heated in the IHX to deliver the needed power to the interfaced HyS equipment. Sand is then stored in the cold sand storage tank at temperatures of the order of 600 °C. It is then lifted up (Bucket Lift in **Figure 2**) to the receiver to complete the heat transfer loop. The IHX and the secondary intermediate loop operate continuously, assuring working continuity for the thermochemical plant, while the solar system and sand receiver operate intermittently.

The HyS plant was designed based on results obtained from a new chemical flowsheet created with an ASPEN Plus simulation for the NHI program. More details and discussions on the new SRNL's HyS flowsheet can be found elsewhere (3). The solar HyS plant, for both the 2015 and 2025 year, was conceptually designed suitably adapting and revising results obtained for the nuclear case to address the solar case working conditions.

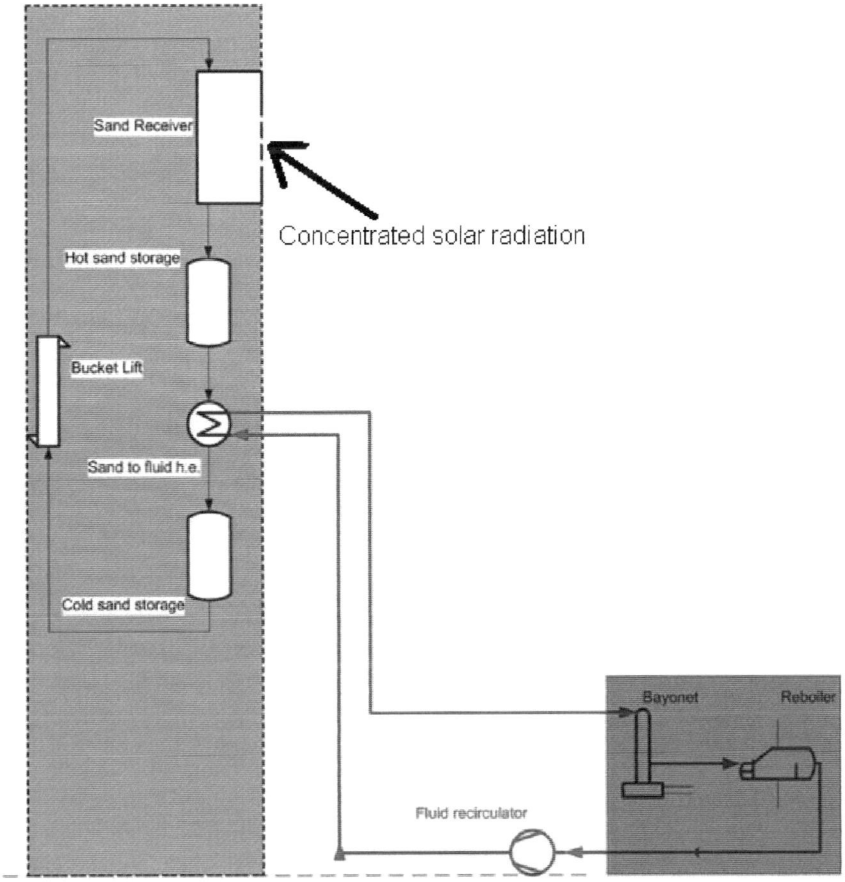

Figure 2: Coupling scheme between HyS plant (interfaced equipment) and solar tower plant based on the particle receiver concept

The 2015 year design assumed reasonable design performance and capital cost estimates that could be expected to be achieved in that time frame. Additional advancements and specific considerations were carried out for the 2025 scenario accounting for benefits coming from potential, but nevertheless realistic, improvements of performances of the overall solar thermochemical plant. For example, the 2015 design assumed two solar central receivers rated at 450 MW maximum thermal input, whereas the 2025 design assumed a single, 700 MW, receiver. Also, the sand-to-gas IHX assumed for the 2015 plant design was removed in the 2025 case and replaced with the interfaced HyS plant equipment directly located inside the solar tower. Regarding the component design of the HyS plant, specific evaluations were carried out concerning the SDE, on the basis of analyses and results obtained at SRNL. The H_2SO_4 decomposition reactor, which is the other challenging and featuring component of the HyS plant, has been designed based on the "bayonet" heat exchanger reactor concept, which has been identified as the baseline concept to decompose H_2SO_4 within the NHI Program (11). The other more traditional components of the thermochemical cycle have been designed considering only commercially available equipment, suitably adapted to the solar working conditions and made of materials capable to resist such highly aggressive environments. A first preliminary conceptual design of the interface system between the solar receiver plant and the HyS process, for the 2015, has also been achieved, based on a sand-to-helium IHX. Further activities are being carried out along with General Atomics to find the best arrangement of the solar interface system, in terms of costs of the device, efficiency and reliability. The solar plant (i.e. heliostat field, solar tower, thermal storage tanks, etc.) has been conceptually designed for both scenarios also considering suggestions and information from Sandia National Laboratory.

Results obtained from the design of the solar plant have been taken as the basis for the economic assessment of the overall process. A standardized cost analysis approach (H2A) developed for the DOE hydrogen program was used as the basis for determining the cost of hydrogen from the plant. SDE installed costs have been evaluated based upon the considerations carried out at SRNL and assuming an installation factor of 1.2, typical for electrolysis systems, which was provided by TIAX, LLC based on other H2A estimates. Regarding the acid decomposition reactor, Westinghouse Electric Co developed and investigated a concept similar to the Sandia Nat Lab bayonet reactor for nuclear applications (12). Costs provided by Westinghouse have suitably been revised and adapted to the solar case conditions, for both the 2015 and the 2025 scenario. Costs of the other (more traditional) equipment of the thermochemical plant have been evaluated using the Aspen Icarus software and suitably revising and (in some cases) re-evaluating those already calculated for the nuclear case. Values so obtained have also been compared with values obtained by available databases giving general agreement. Concerning the solar plant, installed costs have been assessed in conjunction with Sandia National Laboratory and on the basis of values available in literature (13). Indirect costs of the overall solar plant have been estimated following the DOE's H2A guidelines and performing ad hoc evaluations for some specific items (e.g. engineering and design costs for both solar and thermochemical plants). Plant lifetime costs have been assessed considering typical rules for chemical plants and performing specific evaluations for non-traditional aspects (e.g. refurbishment of catalysts for the SDE and H_2SO_4 decomposition reactor). Particular attention was paid to the evaluation of variable operating costs (e.g. electricity cost) which strongly affect the overall H_2 production cost. Finally, the H2A guidelines were followed to assess the H_2 production cost for both time scenarios.

The resulting cost of hydrogen was determined to be approximately 4.80 \$/kgH_2 (US \$ 2005) for the 2025 scenario and 3.19 \$/kgH_2 (US \$ 2005) for the 2025 scenario. A sensitivity analysis and preparation of "tornado charts" were conducted to determine the influence of some quantities and parameters (e.g. efficiency of the HyS plant, costs of the various sections of the solar plant, etc.) and relevant assumptions (e.g. specific costs of the solar field, specific cost of electricity, etc.) on the overall cost for both time scenarios. The economic analysis has been reviewed and approved by TIAX, LLC as part of the work carried out within the STCH project to ensure consistency with other hydrogen production options under consideration.

This technology could be demonstrated at an existing solar test facility, such as the Central Receiver Test Facility at Sandia National Laboratory, in the next few years. Commercialization could occur beginning around 2015. The work has been carried out as part of the DOE-EE Solar Thermochemical Hydrogen (STCH) project.

Bibliography

1. Ivy J. Summary of Electrolytic Hydrogen production. Milestone Report. NREL/MP-560-36734, 2004.
2. Funk JE. Thermochemical hydrogen production: past and present. Int J Hydrogen Energy 2001;26(3):185-190.
3. Gorensek MB, Summers WA. Hybrid Sulfur flowsheets using PEM electrolysis and a bayonet decomposition reactor. Int J Hydrogen Energy (in press).
4. Brown LC, et al. High Efficiency Generation of Hydrogen Fuels Using Nuclear Power. Final Technical Report from General Atomics Corp. to US DOE. GA-A24285, 2003.
5. Farbman GH. The conceptual design of an integrated nuclear-hydrogen production plant using the sulfur cycle water decomposition system. NASA Contr. Report, NASA-CR-134976, 1976.
6. Brecher LE, Spewock S, Warde CJ. The Westinghouse Sulfur Cycle for the thermochemical decomposition of water. Int J Hydrogen Energy 1977;2(1):7-15.
7. Summers WA, et al. Development of a Sulfur Dioxide Depolarized Electrolyzer for the Hybrid Sulfur Thermochemical Process. High Temperature Reactor Conference 2008, Washington, September 30, 2008.
8. Colon-Mercado HR, Hobbs DT. Catalyst evaluation for a sulfur dioxide-depolarized electrolyzer. Electrochemistry communications 2007;9:2649-2653.
9. Summers WA, Corgnale C. Hybrid Sulfur Process Overview for Solar Thermochemical Hydrogen Down-select. DOE-EE Solar Themochemical Hydrogen Down-select Meeting, October 8-9, 2008, Golden, CO, USA.
10. Falcone PK. A handbook for solar central receiver design. Sandia report, SAND 86-8009, 1986
11. Parma EJ, Vernon ME, Gelbard F, Moore RC, Stone HBJ, Pickard PS. Modeling the Sulfuric Acid Decomposition Section for Hydrogen Production. Proceedings of 2007 Int Topical Meeting on Safety and Techn of Nucl Hydrogen Production, Control and Mgmt, Boston, June 24-28, 2007:154-160.
12. Connolly SM, Lahoda EJ, McLaughlin DF. Design of Recuperative Sulfuric Acid Decomposition Reactor for Hydrogen Generation Processes. Proceedings of AIChE 2008 Annual Meeting, Philadelphia, USA, 2008.
13. Sargent and Lundy Group Consulting. Assessment of Parabolic Trough and Power Tower Solar Technology Cost and Performance Forecasts. 2003; SL-5641.

Asbjørn Klerke[1], Jakob Engbæk[2], Rasmus Zink[1] Sørensen, Rasmus Fehrmann[2].

[1] Center for Catalysis and Sustainable Chemistry, Department of Chemistry, Technical University of Denmark.
[2] Danish Technological Institute.

Ammonia and Metal Ammines for high Density Hydrogen Storage

Introduction

The challenge of efficient storage and transportation of hydrogen for use in fuel cells still needs to be adressed. In the following it is described how ammonia has potential to become one of the important indirect hydrogen carriers. Aside from hydrogen, ammonia provides the only carbon-free chemical energy carrier solution, giving no CO_2 emission at the enduser. Ammonia is not a green house gas (GHG), and in light of the global climate challenges with respect to GHG outlined, *e.g.* in the Kyoto protocol, combined with a high hydrogen density, NH_3 could be an interesting alternative to direct hydrogen storage.

Ammonia is also interesting because the infrastructure for production, storage, and transportation, is well known technology from the fertilizer industry. The synthesis of ammonia in the Haber-Bosch process is one of the largest chemical processes with a yearly production of 120 million tons, but this is only enough to support a minor fraction of the hydrogen that will be needed in a future hydrogen economy. Ammonia is primarily produced from natural gas, but hydrogen for the synthesis could be supplied from wind or solar power via electrolysis or any other green source of hydrogen. The synthesis of ammonia is also a prime candidate for CO_2 sequestration to prevent emission of CO_2, because the synthesis gas is already cleaned for CO_2 by pressure swing absorption. Furthermore ammonia plants are typically located near natural gas sites where the CO_2 can be pumped back into the ground. Bulk ammonia is transported and stored as a liquid under pressure or cooled to -33ºC.

Indirect Hydrogen Storage in Ammonia and Metal Ammines

Ammonia as energy storage medium, either for direct use in a combustion engine or indirectly by decomposition to hydrogen and nitrogen, is not new. Several comparative studies of ammonia and other fuels have been made [1,2]. The main conclusion in these studies is that ammonia is a useful alternative to hydrocarbons. Furthermore ammonia is competitive with hydrocarbons in price pr. unit of energy.

Ammonia is a good medium for hydrogen storage on both weight and volume as seen in figure 1. For use in portable devices ammonia can be stored as a high density solid in a range of metal ammine salts. The metal ammines lower the ammonia vapor pressure making them safe to handle and use. In figure 1, mass and volumetric hydrogen densities for storage of 10 kg hydrogen reversibly by 8 different methods is shown[3]. It can be seen that the metal ammines have higher volumetric hydrogen densities than liquid ammonia at room temperature.

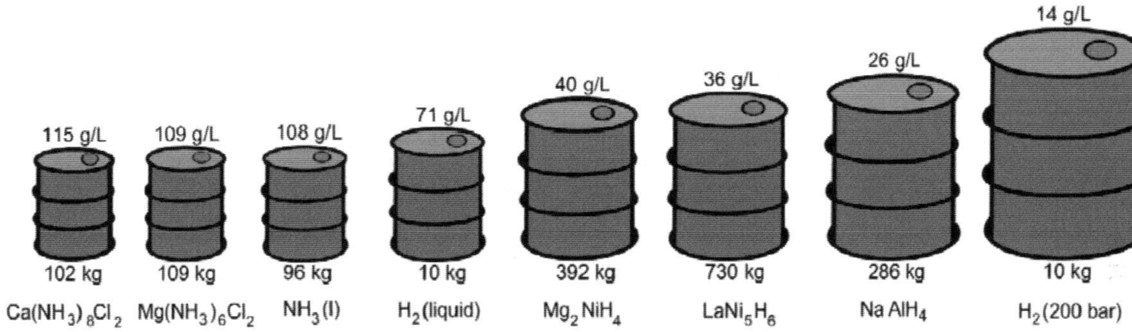

Figure 1: Mass and volume densities of 10 kg hydrogen stored reversibly by 8 different methods, based on the best obtained reversible hydrogen densities reported in the literature without considering the space or weight needed for the container[3].

Desorption of ammonia from metal ammines is easy and only limited by heat transportation[4]. This has been showed using temperature programmed desorption (TPD) measurements. Examples of TPDs can be seen in Figure 2. This also shows that the desorption characteristics are similar for the four investigated salts. The largest variation is the onset temperature of desorption which is determined by the enthalpy. Furthermore the metal ammines can be efficiently compressed and thereby obtain high volumetric hydrogen densities. This can be seen in table 1.

Table 1: Tablet densities and their volumetric and gravimetric hydrogen content[4].

	ρ_{tablet} g/cm^3	% of $\rho_{crystal}$	Volumetric H kgH/L	Gravimetric H wt%H
$Mg(NH_3)_6Cl_2$	1.19	95	0.11	9.19
$Ca(NH_3)_8Cl_2$	1.18	99	0.12	9.78
$Mn(NH_3)_6Cl_2$	1.34	95	0.11	7.96
$Ni(NH_3)_6Cl_2$	1.41	95	0.11	7.83

The densities of the metal ammine chloride tablets are reported and compared to the crystal densities. All the tested metal ammine halides have tablet densities of at least 95% of their crystal densities. The metal ammines can be compressed into a solid medium and still maintain a high desorption rate. This is due to the formation of nanopores during desorption of ammonia[5].

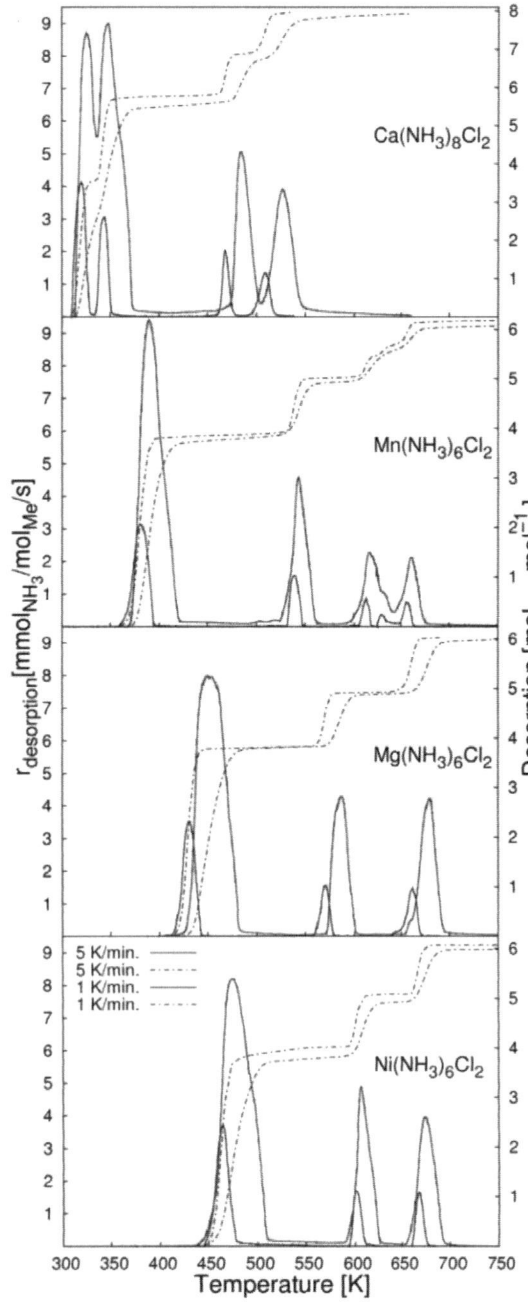

Figure 2: TPDs of $Ca(NH_3)_8Cl_2$, $Mn(NH_3)_6Cl_2$, $Mg(NH_3)_6Cl_2$, and Ni(NH3)6Cl2, with a temperature ramp of 5K/min (red) and 1K/min (blue). The solid lines show the desorption rate and the dashed lines show the amount of ammonia desorped pr metal [4].

Desorption characteristics for the four metal ammine chlorides during TPD are shown in figure 2. As expected from literature data, the trend in temperature for the first desorption peak is Ca<Mn<Mg<Ni. During TPD, $Ca(NH_3)_nCl_2$ has stable compositions with n = 8, 4, 2, 1 and essentially all ammonia desorbs in the temperature range 300-550 K. $Mn(NH_3)_nCl_2$ has stable compositions with n = 6, 2, 1 and two compositions with n<1. Stable structures with less than 1 ammonia molecule pr metal atom are well known for a range of transition metal complexes, and are thus not unlikely even though they have not been reported previously for $MnCl_2$. Essentially all ammonia desorbs in the temperature range 350-675 K. TPDs of $Mg(NH_3)_nCl_2$ were also reported previously and have stable compositions with n = 6, 2, 1. Essentially all ammonia desorbs in the temperature range 410-700 K. During TPD, $Ni(NH_3)_nCl_2$ has stable compositions with n = 6, 2, 1 and all ammonia desorbs in the temperature range 440-700 K. Independently of the temperature ramp, all the TPD peaks start within few degrees of the temperature for 1 bar equilibrium pressure calculated from literature data[6]. This and the near exponential rise in desorption rate with temperature indicates that desorption is equilibrium-limited even at high desorption rates.

The number of metal ammines investigated for use as hydrogen storage materials is still quite small, but due to their similar chemical nature, the general findings can be used as a guideline for the behavior of other metal ammines. This has been tested by detailed modeling using density functional theory (DFT) calculations, which showed good agreement between theoretical calculations and experimental data[7]. DFT calculations are able to reproduce the experimentally observed trend in desorption enthalpies for various ammines at all decomposition steps[4], making such calculations is a valuable tool in the design and prediction of novel metal ammine compounds with specific properties, *e.g.* by varying the metal or halide components in the ammine salt.

Ammonia Decomposition

The ammonia released from metal ammines, can be used directly in a solid oxide fuel cell or a direct ammonia fuel cell. In both types of fuel cells the ammonia is reformed on the anode to form nitrogen and water by reaction with hydroxide in alkaline type fuel cells or by reaction with oxide ions in SOFC and other oxide conducting fuel cells.

The development of new fuel cells is fast but so far the commercialization of new designs is slow. Demo plants with SOFC are operating on different reformed fuel feeds, but no larger tests with ammonia have been published. Therefore the best alternative is to make a separate unit for decomposition of ammonia to hydrogen and nitrogen. This feed can then be used in low temperature fuel cells like PEMFC or PAFC.

The research on ammonia decomposition catalysts show that ruthenium nano-particles supported on high surface area graphite (HSAG) and promoted with cesium is the most active catalyst[8,9,10].

The ammonia decomposition unit is designed for a hydrogen production for a 500-1000W PEMFC. This power demand is chosen as a practical size for a prototype, large enough to have efficient insulation and heat management, and small enough to handle in the laboratory in standard equipment. The power can be used for a small vehicle or as a power back up. This power demand can be covered by decomposition of 4-9 L/min ammonia.

The decomposition of ammonia to hydrogen is endothermic and the reaction can be seen in scheme 1. The heat for the decomposition is supplied by catalytic combustion of ammonia over commercial platinum on alumina catalyst, the reaction is shown in scheme 2.

$2NH_3 \rightarrow N_2 + 3H_2$ $\Delta H=92$ kJ/mol (1)

$2NH_3 + 3/2O_2 \rightarrow N_2 + 3H_2O$ $\Delta H=-633$ kJ/mol (2)

From the enthalpies it can be calculated that catalytic combustion of one mole ammonia develops enough energy for decomposition of 6.8 mole ammonia. With this limit of the system a minimum of 0.73 L/min. ammonia need to be combusted to decompose 5 L/min ammonia. This ratio is not possible to achieve in practice due to heat loss and the need for high temperatures (800-900K) for the decomposition reaction to reach an equilibrium conversion of over 99.8%. This makes heat exchange very important and still energy will be lost as low temperature heat. Some of the heat in the exit gas can be used to desorp ammonia stored in a metal ammine salt and thereby increase the overall energy efficiency of the system. The basic schematic of the ammonia decomposition unit can be seen in Figure 3.

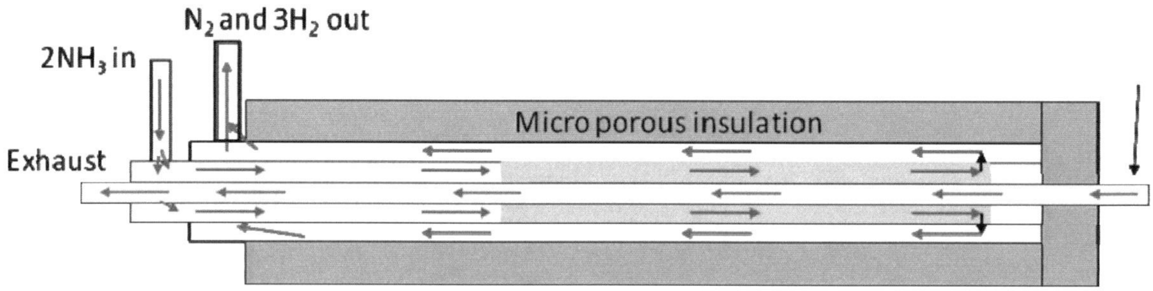

Figure 3: Schematic of ammonia decomposition unit, middle (gray arrows)catalyst bed is Pt on Al_2O_3 for catalytic combustion of ammonia, green bed is Ru on high surface area graphite (HSAG) promoted with Cs for catalytic ammonia decomposition

The design shown in Figure 3 is selected to minimize the energy loss. This is done by placing the ammonia combustion catalyst inside the double pipe, so as much heat as possible can be transferred to the decomposition unit. With this design it is possible to decompose 4.5-5 l/min ammonia by combusting 1L/min. This is an energy efficiency of 74%. If the decomposition unit is build together with a metal ammine storage system for ammonia the hot exhaust can be used to desorp ammonia.

The flow control to the combustion catalyst is important to avoid harmful emission of NO_X and N_2O. They are formed primarily if the ammonia air mixture is oxygen rich. To measure the oxygen amount in the combustion mixture a lambda sensor was used. Exhaust gas samples were taken out at different lambda values around one to verify that a lambda close to one or 450 mV is optimal also for the catalytic combustion of ammonia. The gas samples are investigated with FTIR to determine the amount of slip ammonia and harmful emissions. The results can be seen in Figure 4.

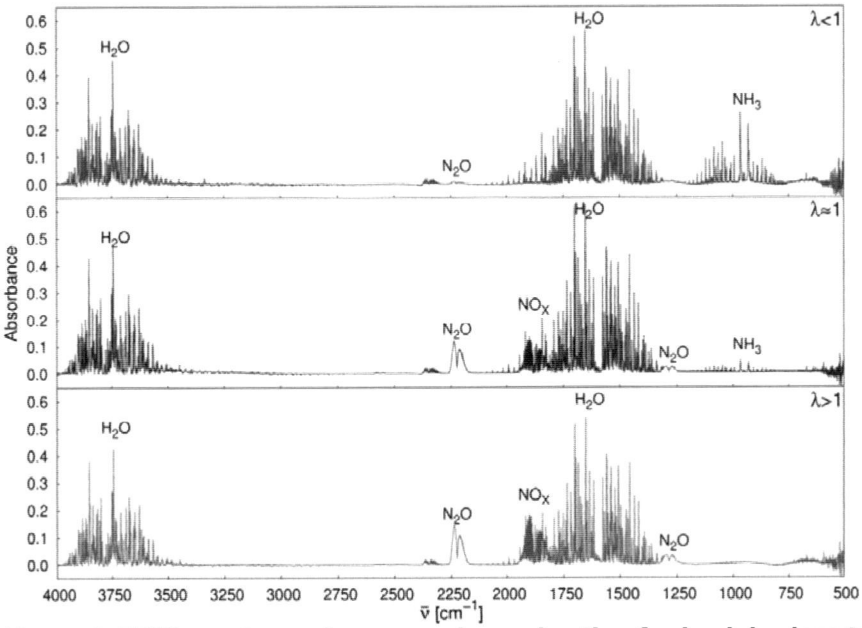

Figure 4: FTIR spectrums for ammonia combustion for lambda close to one.

From the FTIR spectrums in Figure 4, it can be seen that if the lambda value is below one, which mean low oxygen to ammonia ratio, that trace amounts of N_2O can be seen and no NO_X, but not all the ammonia is combusted. The data is as expected as long as there are a surplus ammonia, NO_X formation is depressed but there are still traces of N_2O, this can be seen in the top spectra and lambda below one. At lambda close to one, only trace amounts of ammonia is left and there is a large increase in N_2O and starting formation of NO_X. Finally when lambda is above one and the mixture is oxygen rich, all ammonia is combusted but both formation of N_2O and NO_X increase. The problem with formation of N_2O and NO_X could be avoided with another catalyst.

The tradeoff between not combusting all the ammonia and formation of harmful emissions of N_2O and NO_X, will probably favor not combusting all the ammonia. The main reason for this is that N_2O is a green house gas with a long life time in the atmosphere and NO_X causes smog in urban environments. Ammonia on the other hand is relatively harmless and reacts with the soil as fertilizer. Ammonia can also easily be captured in a solid acid filter, this could be sulphated carbon.

The catalyst for ammonia decomposition is ruthenium promoted with cesium and supported on high surface area graphite (HSAG). This is a very active catalyst and can push the conversion close to the theoretical equilibrium conversion, given by the thermodynamic Gibbs energy. As prototype the catalyst works as expected and gives high conversion, but there are two issues that need to be dealt with mainly coming from the HSAG support. The first is pressure drop duet to, the fine structure of the graphite support and difficulties in making a suitable particle size gives a large pressure drop over the catalyst bed. The second problem is methanation of the graphite, this is due to the fact that ruthenium is a methanation catalyst and that the inside of the catalyst gets very hot (1000K). After the decomposition the reformed gas needs cleaning of slip ammonia. This is done in a sulphated carbon filter. Finally the reformed and clean gas can be fed to a fuel cell. In Figure 5 the reformed ammonia gas is investigated with FTIR to determine the efficiency of the reformer and filter.

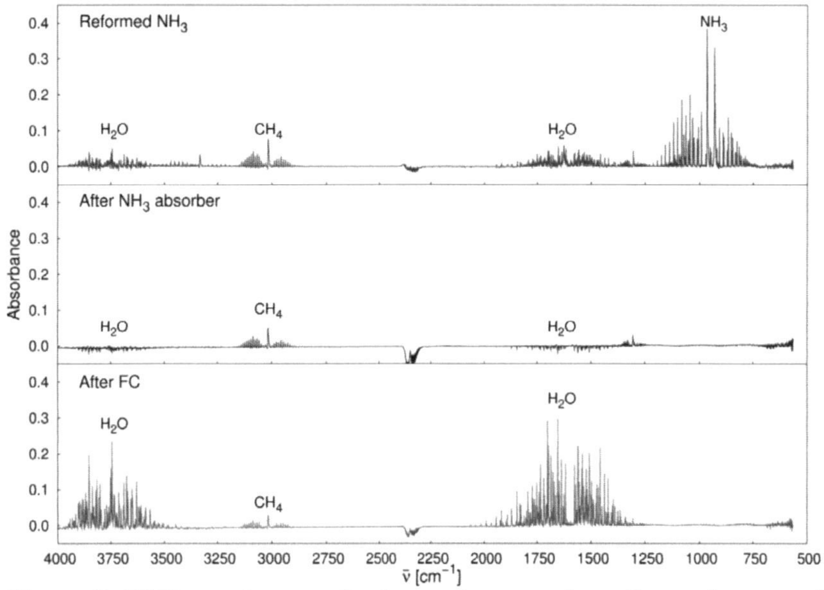

Figure 5: FTIR spectrums of reformed ammonia, reformed ammonia after ammonia slip filter, and finally after the fuel cell.

From the first two spectra in Figure 5 it can be seen that both ammonia and water is removed from the reformed ammonia when it is passed through the sulphated carbon filter. It can also be seen that almost identical amounts of methane is found. The methane is produced from the support material and therefore it is planned to change catalyst support to $MgAl_2O_4$-spinel. This support can handle high temperatures and at the same time it should be easier to lower the pressure drop over the catalyst bed. The last spectrum after the fuel cell is a small experiment, where a PEMFC is feed with the reformed and clean ammonia and as expected the only by product is water.

The data from the experiment of running a fuel cell on the feed of reformed ammonia cleaned in a sulphated carbon filter can be seen in Figure 6. Here a fuel cell is run first on hydrogen and then the gas is changed to the reformed ammonia. All the experiments are at constant current and with the same cell. The change in voltage over time on different feeds is most likely the drying of the membrane, but this is pure speculation. The data is first try at making power directly from the cleaned ammonia reformat.

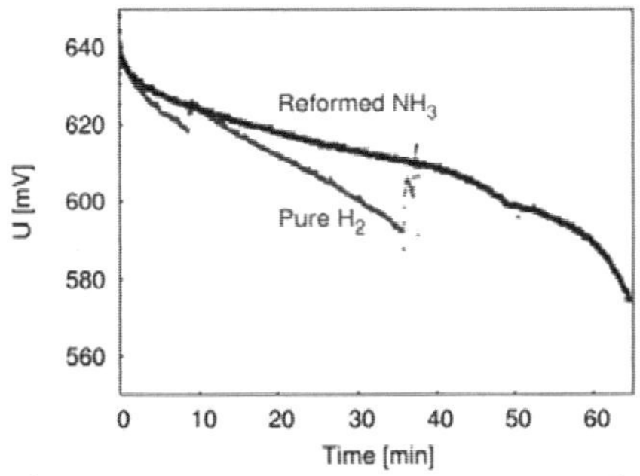

Figure 6: Constant current experiments, with a fuel cell fed with hydrogen and reformed ammonia.

The experiment is done to demonstrate that an ordinary fuel cell can run on the reformed ammonia out of the box. The drop in voltage over time is probably due to drying out of the membrane. The fuel cell was a test cell supplied by IRD fuel cells.

Conclusion

This overview has demonstrated the possibilities of using ammonia as an energy carrier in the hydrogen economy. It has been demonstrated that an efficient ammonia decomposer with high efficiency can be constructed. It has been shown that metal ammines can be used as a solid ammonia storage medium avoiding the hazards of handling and transporting liquid ammonia under pressure. Finally it has been shown that a fuel cell can run directly on the cleaned ammonia reformate. There are still some challenges involved with designing a complete system but the different components are being developed. This involves integration of an ammonia storage unit based on metal ammines, with the developed ammonia decomposition unit and finally connecting it to a full size fuel cell stack.

Acknowledgements

The project is supported by The Danish Council for Strategic Research (Project no. 2104-05-0016). The project is also supported by Signe Shim and IRD Fuel Cells, who supplied the fuel cell.

References:

[1] W. H. Avery, Int. J. Hydrogen Energy, 12 (1988) 761.
[2] R. Metkemeijer, P. Achard, Int. J. Hydrogen Energy, 19 (1994) 535.
[3] A. Klerke, C. H. Christensen, J. K. Nørskov and T. Vegge, J. Mater. Chem. 18 (2008) 2304.
[4] R. Z. Sørensen, J. S. Hummelshøj, A. Klerke, J. B. Reeves, T. Vegge, J. K. Nørskov and C. H. Christensen, J. Am. Chem. Soc. 130 (2008) 8660.
[5] J. S. Hummelshøj, R. Z. Sørensen, M. Y. Kustova, T. Johannesen, J. K. Nørskov, C. H. Christensen, *J. Am. Chem. Soc.*, *128* (2006) 16.
[6] E. Lepinasse, B. Spinner, *Rev. Int. Froid., 17* (1994) 309.
[7] T. Vegge, R. Z. Sørensen, A. Klerke, J. S. Hummelshøj, J. K. Nørskov, C. H. Christensen, *Solid-state hydrogen storage: materials and chemistry*, Woodhead Publishing Limited, 2008.
[8] W. Raróg-Pilecka, D. Szmigeel, Z. Kowalczyk, S. Jodis and J. Zielinski, J. Catal. 218 (2003) 465.
[9] R.Z. Sørensen, L.J.E. Nielsen, S. Jensen, O. Hansen, T. Johannesen, U. Quaade and C.H. Christensen, Catal. Commun. 6 (2003) 229.
[10] R.Z. Sørensen, A. Klerke, U. Quaade, S. Jensen, O. Hansen and C. H. Christensen, Catal. Lett. 112 (2006) 77.

REFORMING OF JP-8 LOGISTICAL FUEL IN A MICROPLASMA FOR COMPACT SOFC POWER: Microfabrication Improvements of Reactors

Peter J. Lindner and R.S. Besser, Department of Chemical Engineering and Materials Science, Stevens Institute of Technology, Hoboken, NJ 07030
201-216-5257; rbesser@stevens.edu

Background

Fuels Cells are attractive for energy production due to the environmental and efficiency problems of combustion processes. The major limitation of hydrogen fuel cell systems has been the lack of a readily available stream of hydrogen. Hydrogen is regarded as impractical for storage hence it will likely be produced by reforming hydrocarbons especially in portable applications. Currently, traditional catalyst processes are used for reforming, but problems due to high device weights, catalyst poisoning issues, long start up time and excessive costs often make them impractical for portable systems [1-3]. To circumvent these issues we are pursuing the use of a non-thermal microplasmas for hydrocarbon reforming.

Currently portable power sources for military systems are supplied through battery power, which in large quantities can become too burdensome for the soldier and the production demands to supply these batteries cannot always be met. Reforming JP-8 logistical fuel with microplasmas would facilitate a long-life, efficient, portable, reusable "fuel processor-fuel cell" system, reducing the dependence of battery technology in the military.

Through a series of atmospheric-pressure microplasma experiments incorporating volatile hydrocarbon species, we have established reproducible plasma operating conditions for a variety of device geometries. These results have led to a next-generation microplasma reactor configuration incorporating favorable geometry and sealed microfluidic flow channels. These improvements are expected to lead to a succession of reforming reaction results in the next round of experiments.

Introduction

The advantages of plasma reforming over catalyst reforming have motivated several research efforts in recent years. There have been various experiments published in literature using plasma as a reforming media for hydrocarbons like diesel, oil, gasoline, JP-8, methane and natural gas among others [1, 2, 4]. These investigations have shown the promise of plasma reforming, but have experienced issues achieving suitable system efficiencies. This project intends to show that microplasmas have advantages in comparison to catalyst, and offer overall improved system efficiency in comparison to conventionally-sized plasmas, both of which promote the future feasibility of microplasma reforming technologies. This overview introduces the characteristics of plasma as a form of matter, the application of plasma as a reforming technology, and the theoretical benefits of non-thermal and micro-scale plasma operation.

Plasma, the fourth state of matter, is defined as a neutral collection of ionized gas species and is the most abundant form of matter in the universe. Some examples include the sun and stars, the aurora borealis, lightning, fire and other such natural phenomena. There are also examples of commercial plasmas such as plasma TVs and fluorescent lights. In industry plasmas have been used for arc welders and plasma torches [5, 6]. Plasma technology is also used commonly to etch and produce small features prevalent in semiconductor fabrication.

This research specifically aims to look at the ability of a plasma to activate a chemical reaction. When comparing plasma processes to catalyst reforming it is important to be aware of the current limitations of catalyst reforming. The three major restrictions to catalysts are there sensitivity to sulfur compounds, the adverse effects of carbon deposits due to coking, and a long start-up time. Sulfur compounds cannot be tolerated at even minuscule concentrations as it will poison the catalyst by blocking active sites. In particular, JP-8 fuel is known to contain sulfur levels up to 5000 ppmw [3, 7]. In order to use catalytic reforming for JP-8, high operating temperatures are required exceeding 800°C, and the catalyst would still be susceptible to poisoning from a monolayer of sulfur [3]. To avoid poisoning the catalytic reformer, the fuel requires a separate hydrodesulphurization (HDS) reaction. The HDS reaction converts sulfur compounds into hydrogen sulfide (H_2S). H_2S is then removed using a zinc oxide (ZnO) absorber. The addition of the HDS step not only complicates the system, but also necessitates an additional feed stream of hydrogen shown in Figure 2.

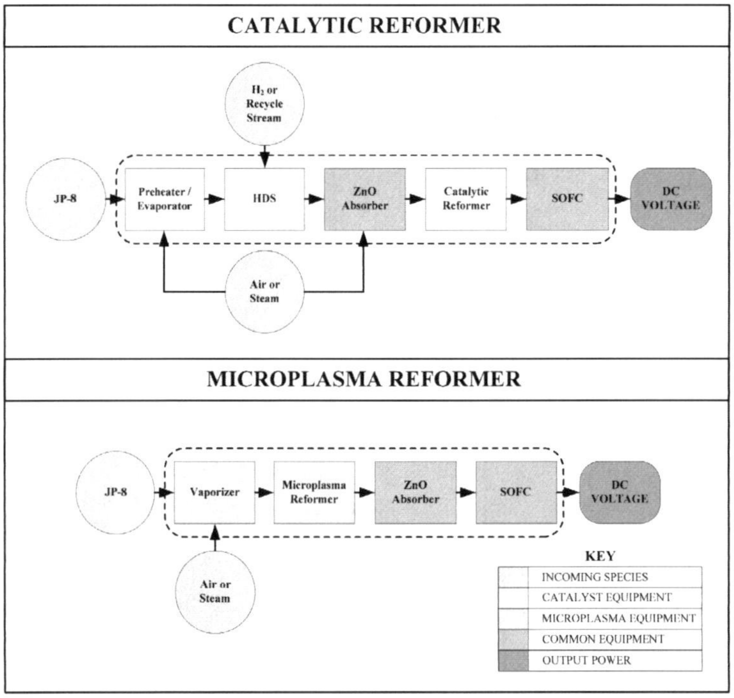

Figure 2. Comparison of Catalyst Reforming and Microplasma Reforming [8]

An additional problem with catalyst reforming is coking. Coking is the deposition of solid porous carbon or ash. The carbon deposits which cause coking are frequent when breaking down large carbon chains. Coke deposits result in the reduction of available active sites leading to overall catalyst deactivation. Coking renders catalytic reforming ill-suited for hydrocarbon reforming processes. Furthermore, the start-up time for a catalytic reformer could take several minutes, rendering a power system useless during the entirety of the start-up period [2].

Plasma technology essentially avoids all of these issues. Earlier work demonstrated that plasmas can break down sulfur compounds present in hydrocarbon fuels, eliminating the need for the HDS step. This was observed in work at Stevens Institute of Technology in collaboration with FuelCell Energy Inc. of Danbury, CT [9, 10]. For this experiment NATO F-76 logistical fuel was sent through a non-thermal plasma device and a significant fraction of the organic sulfur components were converted into H_2S. Although coking is not expected to slow the reaction rate, the effect of coking on the microplasma device has not yet been investigated.

Finally since the plasma is controlled through its own power supply, the start up of the device is instantaneous, determined only by the input power. For these reasons, we propose to further explore plasma technology, especially non-thermal micro-scale plasma devices as a potential reforming mechanism for large hydrocarbon fuels.

Plasma technology provides a non-equilibrium environment in which the generated, chemically-reactive free radicals promote chemical conversion and offer interesting prospects for hydrocarbon reforming [2]. The application of plasmas to reforming processes has recently been studied under various conditions but efficiency issues remain. At MITs Plasma Science and Fusion Center (PSFC) a non-thermal plasma reactor called a plasmatron was tested with various fuels. Although very successful in chemical conversion, the device could not produce energy efficiencies beyond 74% [4]. Providing higher electron and free radical densities should theoretically increase the overall efficiency.

A microplasma is a plasma confined to at least one dimension 1 mm or less [11]. As an emerging field of low temperature plasma science and technology microplasmas have much higher power densities (exceeding 1 kW/cm^3), higher electron densities (exceeding 10^{15} cm^{-3}) and increased surface-to-volume ratios when compared to conventional, large-scale plasma-chemical systems. A high surface-to-volume ratio imparts excellent thermal management and mixing characteristics that help maintain homogeneous, isothermal reacting volumes [12]. These microplasma characteristics present processing advantages for hydrocarbon reforming applications. Moreover, conventional plasma processes generally operate at lower pressures allowing the system to meet the Paschen Minimum, minimum breakdown voltage. Table 1 shows the minimum breakdown voltage for various gases and the ideal electrode distance, d, for atmospheric pressure. Processing with microplasmas makes operating at atmospheric pressures possible due to the small distance between electrodes. By running at atmospheric pressure we can minimize equipment requirements and simplify the overall operating system.

Table 1. Minimum Sparking Constants for Selected Gases [13]

Gas	Cathode	$V_{b,\,min}$ (Volts)	pd_{min} (10^{-3} Torr-m)	d_{min} at 760 Torr (μm)
Ag	Fe	265	15.0	19.7
Air	-	360	5.7	7.5
CO_2	-	420	5.0	6.6
H_2	Pt	295	12.5	16.4
He	Fe	150	25.0	32.9
Hg	W	425	18.0	23.7
Hg	Fe	520	20.0	26.3
N_2	Fe	275	7.5	9.9
N_2O	-	418	5.0	6.6
O_2	Fe	450	7.0	9.2
SO_2	-	457	3.3	4.3

Microplasma technology offers many advantages in the devices design. They are generally small and can be designed to be any shape. The devices can easily include micro fluidic connections to allow for a flow through process. Integrated electrodes and heat exchangers for the individual microplasma devices are possible. Scaling up from experimental setup to the final design is unnecessary as the reactors would have the same dimensions. Moreover micro fabrication devices are relatively inexpensive to produce. These advantages make microplasmas ideal for a portable hydrocarbon fuel reforming process.

Research Accomplishments

To date two generations of microplasma reactor chips have been designed and fabricated using silicon microfabrication technology at the Cornell Nanofabrication Facility of Cornell University in Ithaca, New York. The microplasma reactors produced are microhollow cathode (MHC) devices with a variety of geometries. The first generation of miocroplasma reactors were open-faced plasma devices produced to determine the characteristics of the plasmas with various gaseous. These microplasma devices had various chip geometries which were tested to determine the best characteristics for future study. The second generation microplasma devices include flow through connections and a sealed top. This allows for testing the chemical reaction characteristics of the plasma.

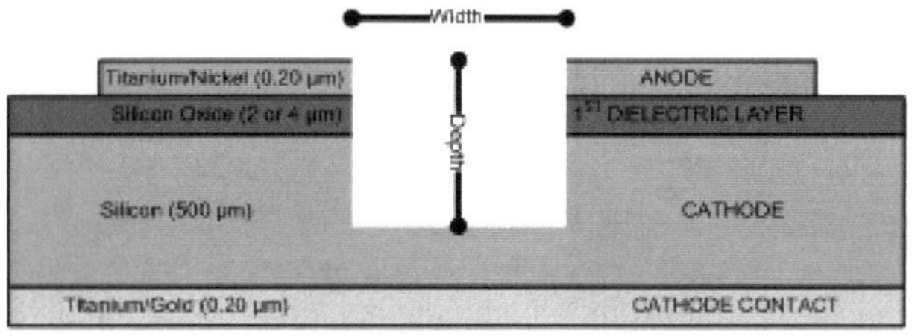

Width	25µm, 100µm, 250µm
Depth	25µm, 100µm, 250µm
Sillcon Oxide	2µm, 4µm
Enhanced Perimeter	Yes or No

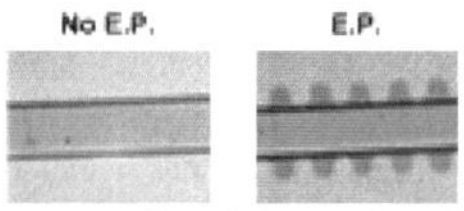

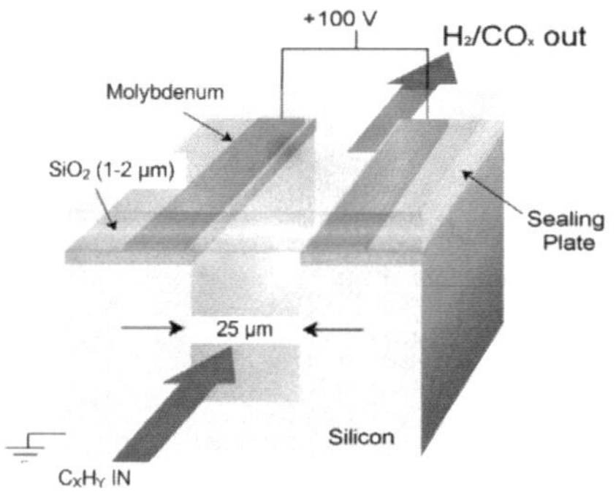

Figure 3. Microhollow Cathode Devices

To test our devices a vacuum chamber system was completed to allow for testing and verification of plasma behavior. Initially tests were done with various inert gases such as helium, neon and nitrogen. To date inert plasma experiments have shown our devices to be comparable to literature. We have also observed the effects of device geometry and the degradation of the devices. Developing butane plasma began our work on volatile hydrocarbons. Butane experiments gave promise to developing a plasma with hydrocarbon fuels. Experiments with butane determine the effects of microplasma device geometries.

Results

To determine the ideal dimensions for our second generation devices butane was tested with various geometries. Tests were done by limiting the applied current while monitoring the current and voltage. These tests sustained point plasmas of butane that exceeding power densities of 10MW/cm^3. Figure 4 shows an example of data from a butane plasma experiment.

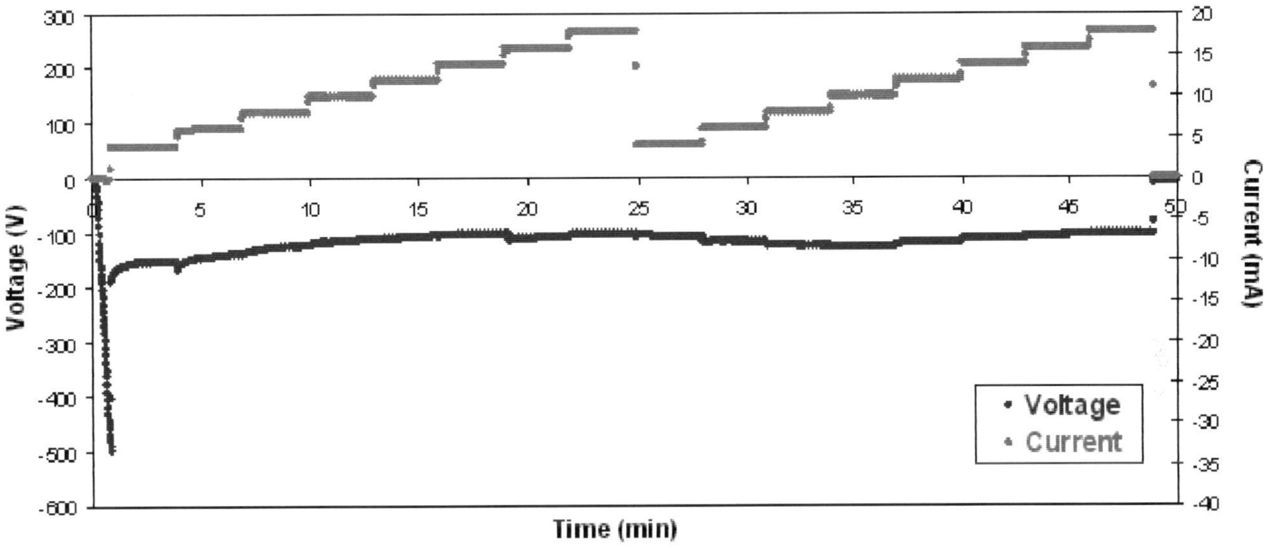

Figure 4. Microplasma Butane at 760 torr

These experiments gave us the ideal dimensions for the second generation micro fluidic reactors.

Currently the second generation micro fluidic plasma devices are being tested to determine the chemical reaction conversion of butane and other various gaseous hydrocarbons. These experiments will move us closer to the final goal of reforming JP-8 to provided portable power on the battlefield. The results of the experiments will be presented at the conference in detail.

Acknowledgements

The authors wish to express their gratitude to the Army Research Office for supporting this project and to various individuals which contributed to the experiments (E. Whittaker, G. Wohlrab, B. Fraser, A. Najem, K. Ciampi, E. Lennon, J. Hwang, A. Omosebi, J. Weinstein, T. Heithoff, S. Hwang, J. Symonds).

References

1. G. Petitpas, J.-D. Rollier, A. Darmon, J. Gonzalez-Aguilar, R. Metkemeijer, L. Fulcheri, "A comparative study of non-thermal plasma assisted reforming technologies," *International Journal of Hydrogen Energy*, **32**, 2848-2867 (2007).
2. T. Paulmier, L. Fulcheri, "Use of non-thermal plasma for hydrocarbon reforming," *Chem. Eng. J.,* **106**, 59-71, (2005).
3. M.V. Mundschau, C. G. Burk, D. A. Gribble Jr., "Diesel fuel reforming using catalytic membrane reactors," *Catalysis Today*, **136**, 190–205, (2008).
4. D. R. Cohn, L. Bromberg, K. Hadidi, "Onboard Plasmatron Hydrogen Production For Improved Vehicles," *MIT Plasma Science and Fusion Center*, (2006).
5. Wikipedia contributors, "Plasma (physics)," *Wikipedia, The Free Encyclopedia*, (2008). Available at: http://en.wikipedia.org/w/index.php?title=Plasma_(physics)&oldid=233794520.
6. Wikipedia contributors, "Fusion Energy," *Wikipedia, The Free Encyclopedia,* (2008). Available at: http://en.wikipedia.org/w/index.php?title=Fusion_power&oldid=233899672.
7. H. Scott Fogler, "Elements of Chemical Reaction Engineering 3rd Edition", *Prentice Hall PTR*, (1999).
8. D. J. Moon, J. W. Ryu, K. S. Yoo, D. J. Sung, S. D. Lee, "Development of iso-octane fuel processor system for fuel cell applications," Catalyst Today, **136**, 222–227, (2008).
9. K. Becker, Final Report, "Compact Plasma Reactor for Ship Service Fuel Cell System," *Office of Naval Research Ship Service Fuel Cell Program*, pg 234, (2004).
10. D. Dietz, H. Ghezel-Ayagh, J. Hunt, A. Belkind, K. Becker, and A. Nickens, "Plasma Treatment of a Heated Diesel/Steam Mixture for Use in Ship Services Fuel Cell Systems", *Proc. 31st IEEE ICOPS*, Baltimore, MD, pg.412, (2004).
11. R. Foest, M. Schmidt, K. Becker, "Microplasmas, an emerging field of low-temperature plasma science and technology," International Journal of Mass Spectrometry, **248**, 87-102, (2006).
12. X. Ouyang, L. Bednarova, P. Ho, and R.S. Besser, "Preferential Oxidation of Carbon Monoxide in a Thin-Film Catalytic Microreactor: Advantages and Limitations," *AIChE J.,* **51** (6), 1758-1772, (2005).
13. J. Reece Roth, "Industrial Plasma Engineering", *Institute of Physics Publishing, Bristol and Philadelphia*, (1995).

55f Coke and Microreactors: a Contradiction? - Catalytic Cracking of Propane in a Microreactor Heated by Catalytic Combustion

Oliver Görke, Peter Pfeifer, and Klaus Schubert. Institut für Mikroverfahrenstechnik, Forschungszentrum Karlsruhe, Hermann-von-Helmholtz-Platz 1, Eggenstein-Leopoldshafen, 76344, Germany

Coke and Microreactors: A Contradiction? - Catalytic Cracking of Propane in a Microreactor Heated by Catalytic Combustion

The cracking of propane to generate hydrogen for fuel cell systems is promising due to the availability of excess C3/C4 fractions from oil refinery and a resulting high hydrogen purity which is a demanded for applications like auxiliary power units, battery chargers or mobile generators. Microchannel reactors offer the potential to cope with sudden hydrogen demand and load changes.

Although it is commonly agreed that coke or soot formation has to be avoided in microreactors, we can demonstrate in this study the successful application of a microchannel reactor to produce hydrogen by propane cracking. The used reactor contains four passages. The reactor operates in transient mode, i.e. two intervals. In passage 1, filled with stainless steel mesh, propane is cracked in interval 1 and hydrogen is formed, but also soot is formed. In passage 2, also filled with a stainless steel mesh, soot is simultaneously oxidized in interval 1. In interval 2, the formed soot is oxidized in passage 1, and propane is cracked in passage 2. Between the two intervals, the former respective cracking passage has been flushed in the experiments by nitrogen for about half a minute for safety reasons. Passage 3 serves as supply of simulated anode-off-gas to passage 4 but it could be also used to supply CO from coke burning. Passage 3 is connected to passage 4 by micro holes to prevent homogeneous reaction. Air is added in passage 4 to perform the catalytic combustion.

By mass spectrometry, the product gas stream in the two intervals was analyzed. It can be shown that propane can successfully be cracked reaching an hydrogen yield of more than 70 % for about 8 minutes. In the second interval the generated soot can be oxidized within 20 minutes in air. The balance shows that the heat of oxidizing soot and of oxidizing anode-off-gas is sufficient to drive the endothermal cracking process. The oxidation of soot can be accelerated by impregnation of the stainless steel mesh with V_2O_5 without decreasing cracking activity.

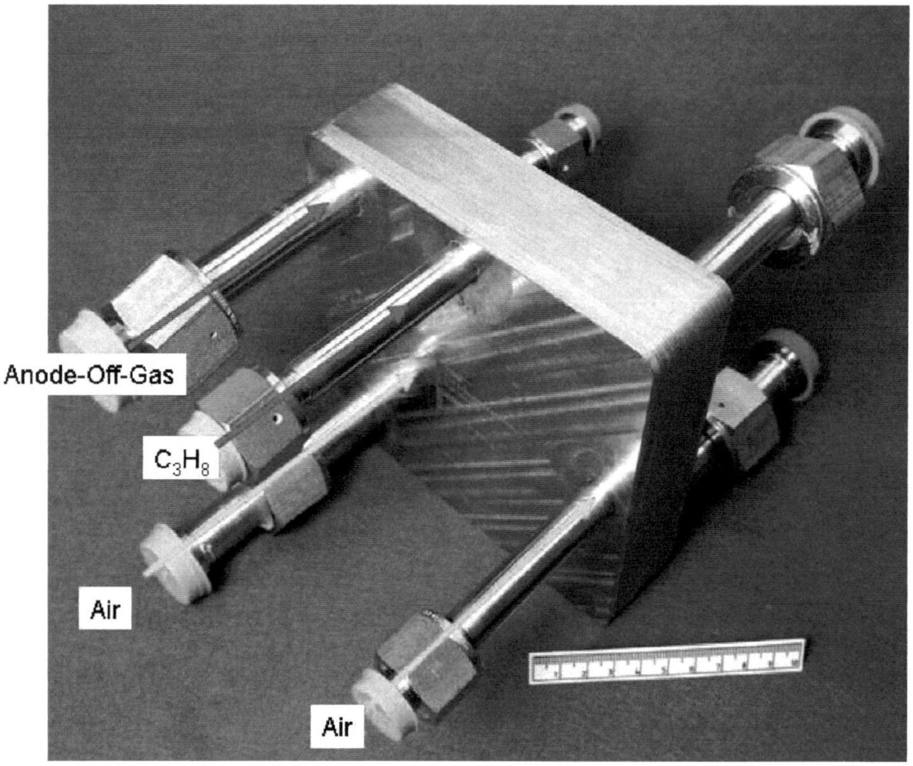

Fig. 1: Propane crack reactor

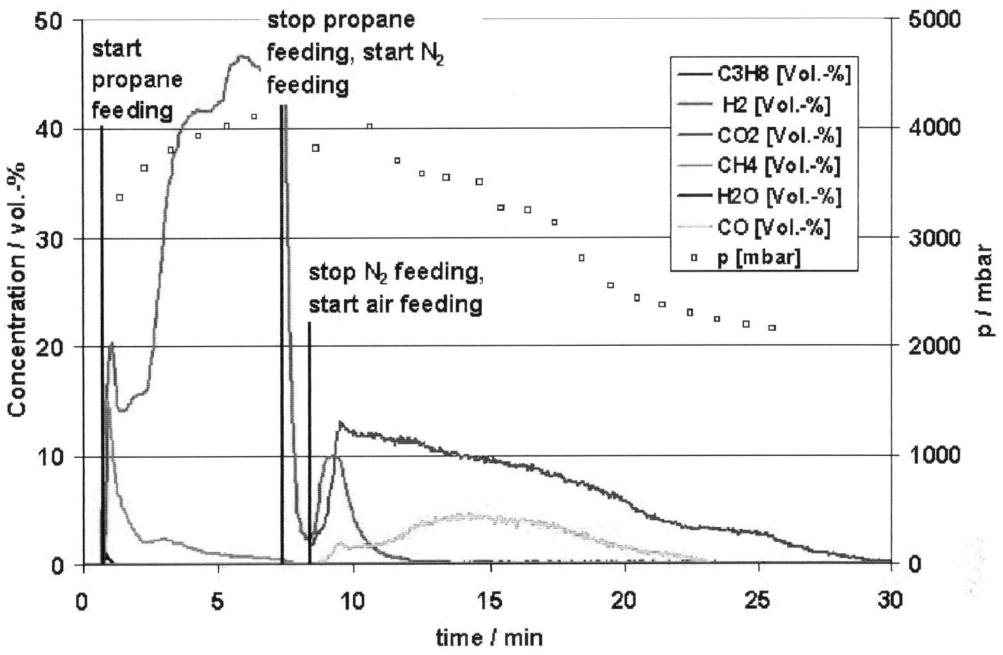

Fig. 2: product gas of passage 1 of one complete cycle

55h The USDOE Hydrogen Program: Status and Performance of Hydrogen Production and Delivery Technologies

Monterey Gardiner, J. Holladay, R. Garland, and R. Farmer. Office of Hydrogen, Fuel Cells & Infrastructure Technologies, U.S. Department of Energy, EE-2H 1000 Independence Ave. SW , Washington, DC 20585

The DOE Hydrogen Program is pursuing a number of diverse pathways for producing hydrogen at both distributed and centralized locations. Projections show that near term production of hydrogen at the fueling station using steam-methane reforming can meet the DOE cost targets of $2.00 - $3.00/gge, untaxed and delivered. The Program's near-term efforts are now focused on distributed production from the reforming of bio-derived liquids, as well as small-scale water electrolysis. Longer term centralized production of hydrogen from coal with sequestration and biomass gasification along with nuclear energy provides additional opportunities and is under investigation. Alternate approaches to these technologies include solar energy for water splitting using photobiological, photoelectrochemical, and thermochemical processes. The objective of hydrogen delivery R&D is to reduce the total cost of delivery to less than $1/gge. Focus areas include the analysis and development of hydrogen compression technology (centrifugal and electrochemical), stationary storage and tube trailers, liquefaction technologies, new pipeline materials, and new liquid or solid hydrogen carriers.

Transport through Partly Saturated Beds of Fully Wetted Hydrophilic Fibers

Manolis M. Tomadakis, Florida Institute of Technology, Melbourne, FL 32901

Liquid-saturated beds of hydrophilic fibers find numerous modern technology applications, including many in new energy and environmental systems, such as titanium dioxide light-irradiated superhydrophilic fibers used in photocatalytic decomposition of airborne pollutants and wastewater remediation,[1] lignocellulosic cotton fibers used in diesel oil sorption from water,[2] and environment friendly composite materials utilizing the hydrophilic fibers in natural polymers, such as starch, protein, and cellulose.[3] Hydrophobic fibers are the material of preference in fuel cell gas diffusion and catalyst layers as they facilitate water removal from the cell stacks. Nonetheless, hydrophilic fibers are also utilized in certain fuel cell applications, since they enhance the dispersion of the catalyst particles, hence also the catalytic activity and fuel cell efficiency,[4] while helping maintain the electrolyte wet in systems operating at low relative humidity conditions.[5] The high hydrophilicity and electrical conductivity of sintered stainless steel fiber felt gas diffusion layers are credited for enhancing the performance of air-breathing direct methanol fuel cells.[6] Not surprisingly, hydrophilic fibrous media have been investigated alongside hydrophobic ones in a number of recent experimental and theoretical fuel cell studies.[7-10]

Our experimentally validated earlier and recent numerical methods for estimating effective mass, energy, and momentum transport properties of various types of fibrous media were modified to account for flow through beds of fully wetted hydrophilic fibers of various porosities and saturation levels. The numerical predictions for the effective diffusivity and viscous permeability of such beds are in agreement with earlier analytical predictions and experimental data of the literature for flow through partly saturated fibrous media and packed beds of cylinders. Further modification of our algorithms is under way, to derive the effective thermal and electrical conductivity and magnetic permeability of such systems.

I. Effective Diffusivity and Viscous Permeability

A. Dry Media

Tomadakis and Sotirchos[11] computed the effective diffusivity, D_e, of various types of random and ordered fiber structures, including the bi-directional (2-d) structures made of fibers randomly placed and oriented on parallel planes, as illustrated in Fig. 1.

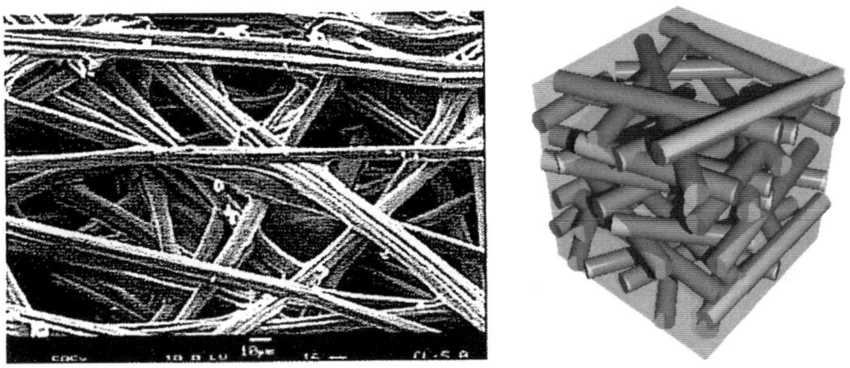

Figure 1: SEM photo of fibrous carbon paper[12] and computer illustration of fiber orientation in numerical samples.[13]

They presented their results in terms of the tortuosity coefficient, η, defined by $D_e = D\varepsilon/\eta$, where ε is the porosity of the structure and D is the reference diffusivity; when considering bulk (molecular) diffusion, the reference diffusivity becomes equal to the bulk self-diffusion coefficient of the fluid. The resulting values of the bulk tortuosity, η^b, were fitted into a modified Archie's Law[11]

$$\eta^b = \left(\frac{1-\varepsilon_p}{\varepsilon-\varepsilon_p}\right)^{\alpha}; \qquad D_e = D\varepsilon\left(\frac{\varepsilon-\varepsilon_p}{1-\varepsilon_p}\right)^{\alpha} \tag{1a,b}$$

with the percolation threshold, ε_p, reported equal to 0.11 for flow through 2-d fiber structures. Archie's exponent, *α,* was found equal to 0.785 for diffusion normal to the fiber planes (cross-plane flow) and 0.521 for diffusion parallel to such planes (in-plane flow). When compared to the predictions of Eq. (1b), a popular simplifying approximation employed in many modeling studies, $D_e = D\varepsilon^{1.5}$, overpredicts the diffusivity by up to a factor of 3 for $\varepsilon > 0.2$, and even more at lower porosities. Vignoles and co-workers[14-15] measured the transport and structural properties of a fibrous carbon-carbon composite preform using x-ray computerized microtomography, and reported excellent agreement with the predictions of Eqs. (1a,b).

Tomadakis and Robertson[13] employed the conduction-based permeability theory by Johnson et al.[16] to obtain the viscous tortuosity, η^v, and viscous permeability, *k*, of random fiber structures from the bulk tortuosity values reported earlier.[11] They derived

$$\eta^v = \eta^b\left(1+\frac{\alpha\varepsilon}{\varepsilon-\varepsilon_p}\right)^2; \qquad \frac{k}{r^2} = \frac{\varepsilon}{8\eta^v \ln^2\varepsilon} = \frac{\varepsilon}{8\ln^2\varepsilon}\left(\frac{\varepsilon-\varepsilon_p}{1-\varepsilon_p}\right)^{\alpha}\left(1+\frac{\alpha\varepsilon}{\varepsilon-\varepsilon_p}\right)^{-2} \tag{2a,b}$$

where *r* is the fiber radius and ε_p, α are the same Archie's law parameters used in Eqs. (1a,b). The predictions of Eqs. (2a,b) are in very good agreement with numerous experimental measurements of the viscous permeability of random fiber structures.[13] Gostick et al.[17] compared the predictions of Eqs. (2a,b) to their own experimental measurements of both in-plane and through-plane permeability of many types of fibrous fuel cell diffusion media, and reported very good agreement as well. Tomadakis and Pentas[18] employed Eqs. (2a,b) in their PEM Fuel Cell modeling study and found that the fibrous gas diffusion layer (GDL) permeability values routinely used in the literature may lead to significant errors in the fuel cell modeling predictions, overestimating the current and power density by 100% or more.

B. Fluid Saturated Media

In partly saturated porous media, the presence of liquid introduces additional resistance to mass transport by reducing the area available for diffusion and increasing the tortuosity of the gas diffusion paths. The resulting reduced diffusivity of the fluid-saturated medium, D_e^s, is related to that of the dry medium, D_e, through the normalized relative diffusivity function $g(s)$

$$g(s) = D_e^s / D_e = (1-s)^m \tag{3}$$

s being the saturation parameter (fraction of pore space occupied by the liquid: $s = \varepsilon_\ell/\varepsilon$) and *m* an adjustable exponent. A practically second-order dependence of *g(s)* on *(1-s)* was reported by Martys[19] in his numerical study of the effects of saturation on the diffusivity of porous media with spherical incursions (for the ideal case of one fluid only wetting the whole porous surface, while the other fluid assumed non-wetting). An exponent value m=1.5 is predicted by the effective medium theory for systems where the pore geometry remains constant, unaffected by fluid saturation.[20] The significant D_e^s reduction at high saturation levels predicted by Eq. (3) is in line with experimental results of the literature.[20]

Just as the effective diffusivity, the gas-phase permeability of fluid-saturated porous media, k_g^s, is described using a normalized function, namely the relative permeability $k_{rg}(s)$

$$k_{rg}(s) = k_g^s / k_g \tag{4}$$

where k_g is the gas-phase permeability of the dry medium. Although k_{rg} depends on other parameters too, such as the pore geometry, contact angle, surface tension, and density ratio of the involved fluids,[21] its saturation dependence has attracted the most research interest. It has been investigated in a number of studies for various types of porous media and gas-liquid or liquid-liquid systems; the results were fitted into $k_{rg}(s)$ correlations of various forms,[21] including the simple

$$k_{rg}(s) = (1-s)^m \tag{5}$$

where the value of the fitting exponent m is determined experimentally (and is different, in general, than that for the relative diffusivity, Eq. (3)). Saez and Carbonell[22] processed many experimental $k_{rg}(s)$ data points from earlier studies (obtained for air-water flow through various particle beds), including the results by Specchia and Baldi[23] and Larkins[24] for flow through packed beds of cylinders. They fitted all data into Eq. (5) and reported that an exponent value m=4.8 describes adequately all data sets. Levec et al.[25] investigated air-water flow through packed beds of spheres, and reported Eq. (5) m exponents in the range 3.6-5.3 for low gas-phase Reynolds numbers, i.e., when viscous flow conditions were approached.

Core-Sheath Model

We propose to model partly saturated beds of fully wetted hydrophilic fibers by means of a core-sheath model. According to that postulated topology, each cylindrical fiber of radius r is surrounded by an annular liquid sheath of thickness δ_ℓ, which wets completely the solid surface, separating it from the gas phase. The total porosity and solid fiber volume fractions of the resulting structures, ε and ε_f, the fiber surface area per unit volume, S_f, and the mean intercept length of the solid phase, d_f (a characteristic size parameter) are obtained from[26]

$$\varepsilon = 1-\varepsilon_f = \exp(-\pi l r^2)\,; \qquad S_f r = -2\varepsilon \ln\varepsilon\,; \qquad \frac{d_f}{r} = \frac{4\varepsilon_f}{S_f r} = \frac{-2\varepsilon_f}{\varepsilon \ln\varepsilon} \tag{6a-c}$$

where l is the length of fiber axes per unit volume of the bed. The fluid phase volume fraction, ε, is distributed between the liquid and gas phases at each saturation level, s, according to

$$\varepsilon_\ell = \varepsilon s\,; \qquad \varepsilon_g = \varepsilon - \varepsilon_\ell = \varepsilon(1-s) \tag{7a,b}$$

The thickness of the liquid sheath surrounding each fiber, $\delta_\ell = r_\ell - r$, is derived using Eq. (6a) and the equivalent relation for the gas-phase porosity, $\varepsilon_g = \exp(-\pi l r_\ell^2)$

$$\frac{\delta_\ell}{r} = \frac{r_\ell}{r} - 1 = \sqrt{\frac{\ln \varepsilon_g}{\ln \varepsilon}} - 1 \tag{8}$$

The gas-liquid interface area per unit volume, S_g, and the gas and liquid phase mean intercept lengths, d_g, d_ℓ are obtained from

$$S_g r_\ell = -2\varepsilon_g \ln \varepsilon_g; \qquad \frac{d_g}{r_\ell} = \frac{4\varepsilon_g}{S_g r_\ell} = \frac{-2}{\ln \varepsilon_g} \tag{9a,b}$$

$$\frac{d_\ell}{r_\ell} = \frac{4\varepsilon_\ell}{(S_f + S_g) r_\ell} = \frac{-2\varepsilon_\ell}{\varepsilon \ln \varepsilon (r_\ell / r) + \varepsilon_g \ln \varepsilon_g} \tag{10}$$

The proposed core-sheath fiber model enables a straight-forward calculation of the gas-phase effective diffusivity and viscous permeability of the partly saturated bed of hydrophilic fibers. Equations (1b) and (2b) can be used for that purpose, with the total porosity, ε, and fiber radius, r, replaced by the volume fraction of the non-wetting fluid (gas-phase porosity), ε_g, and the liquid sheath radius, r_ℓ, respectively. When the predictions of the updated equations are compared to the corresponding ones for the dry medium (Eqs. (1b) and (2b)), the equivalent relative diffusivity and permeability functions are readily obtained

$$g(s) = \frac{\varepsilon_g}{\varepsilon}\left(\frac{\varepsilon_g - \varepsilon_p}{\varepsilon - \varepsilon_p}\right)^{\alpha} = (1-s)\left(1 - \frac{s}{1 - \varepsilon_p/\varepsilon}\right)^{\alpha} \tag{11}$$

$$k_{rg}(s) = g(s)\frac{\ln \varepsilon}{\ln \varepsilon_g}\left[\left(1 + \frac{\alpha\varepsilon}{\varepsilon - \varepsilon_p}\right)\Big/\left(1 + \frac{\alpha\varepsilon_g}{\varepsilon_g - \varepsilon_p}\right)\right]^2 =$$

$$= g(s)\frac{\ln \varepsilon}{\ln[\varepsilon(1-s)]}\left[\left(1 + \frac{\alpha}{1 - \varepsilon_p/\varepsilon}\right)\Big/\left(1 + \frac{\alpha(1-s)}{1 - s - \varepsilon_p/\varepsilon}\right)\right]^2 \tag{12}$$

Eqs. (6b) and (7a,b) were also used in the derivation, to express ε_g in terms of s and obtain $k_{rg}(s) = k_g^s / k$ from the dim/less permeability ratio $\left(k_g^s / r_\ell^2\right)/\left(k_g / r^2\right)$. The predictions of Eqs. (11) and (12) are presented in Figures 2 and 3, respectively, for cross-plane and in-plane flow through fibrous media, for a broad range of porosity (0.4-0.72) and water saturation (0-0.5). Fitting these predictions to the commonly used exponential forms (Eqs. (3) and (5)) leads to the equivalent *m* values predicted by our model. As illustrated in Fig. 2, these values fall in the range 2-2.2 for cross-plane diffusion and 1.6-1.8 for in-plane, showing only a weak dependence on porosity and saturation level. These predictions agree with the results of Nam and Kaviany,[20] Martys,[19] and the effective medium theory,[20] which predict *m* exponent values for the relative diffusivity of gas diffusion layers and other porous media in the range 1.5-2.

Regarding the *m* values predicted by our model for the relative permeability (Fig. 3), they range from 3.6 to 4.4 for cross-plane flow, and 3.2 to 4 for in-plane flow. These results are in line with the literature values discussed above, where *m* is seen to range from 3.6 to 5.3 for viscous flow through beds of cylinders and other particles.[21-23] The literature values obtained for beds of cylinders[23-24] are also plotted in Fig. 3, for the sake of comparison.

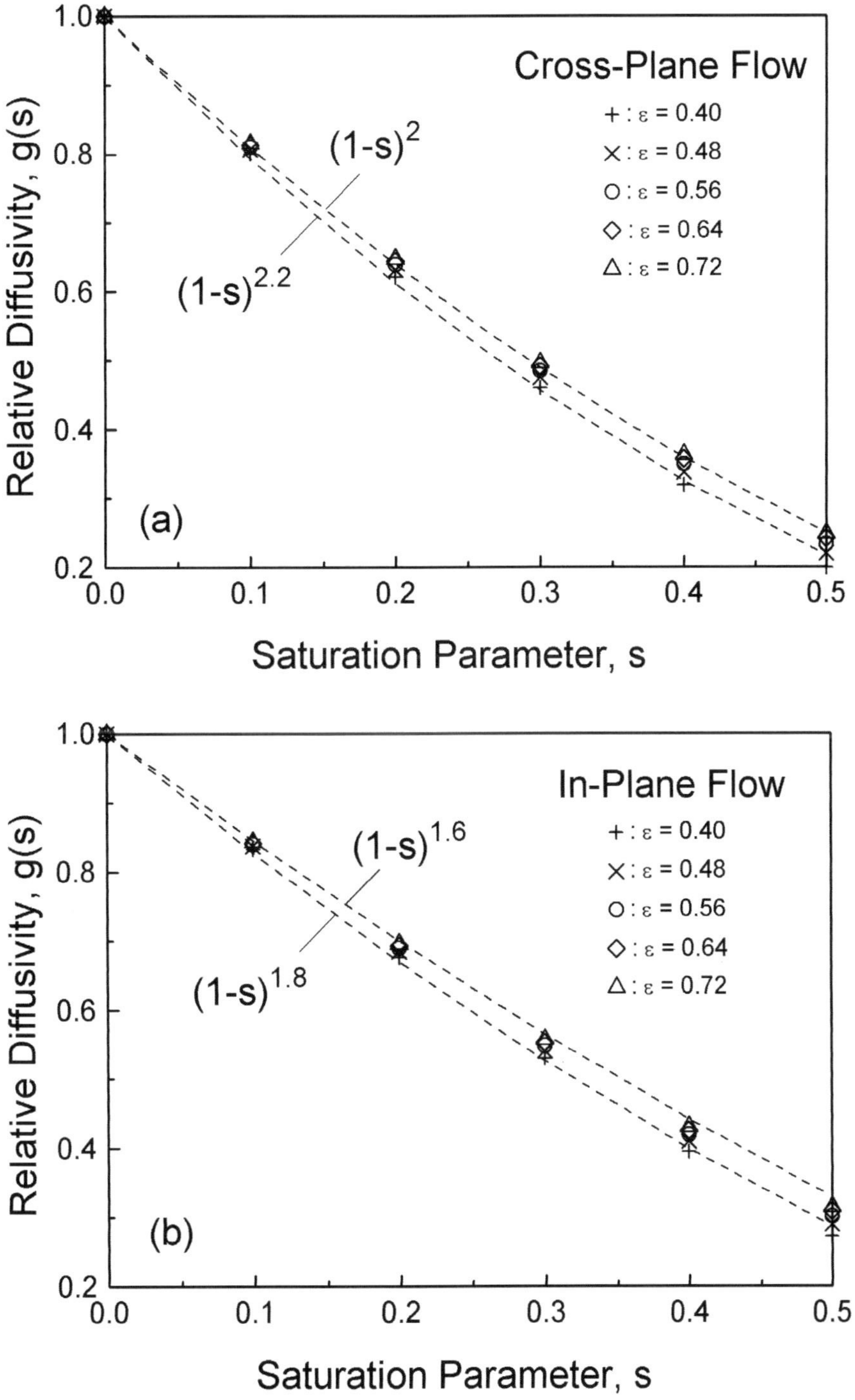

Figure 2. Relative diffusivity predictions of the core-sheath fiber model for (a) cross-plane and (b) in-plane flow through partly saturated beds of 2-d hydrophilic fibers

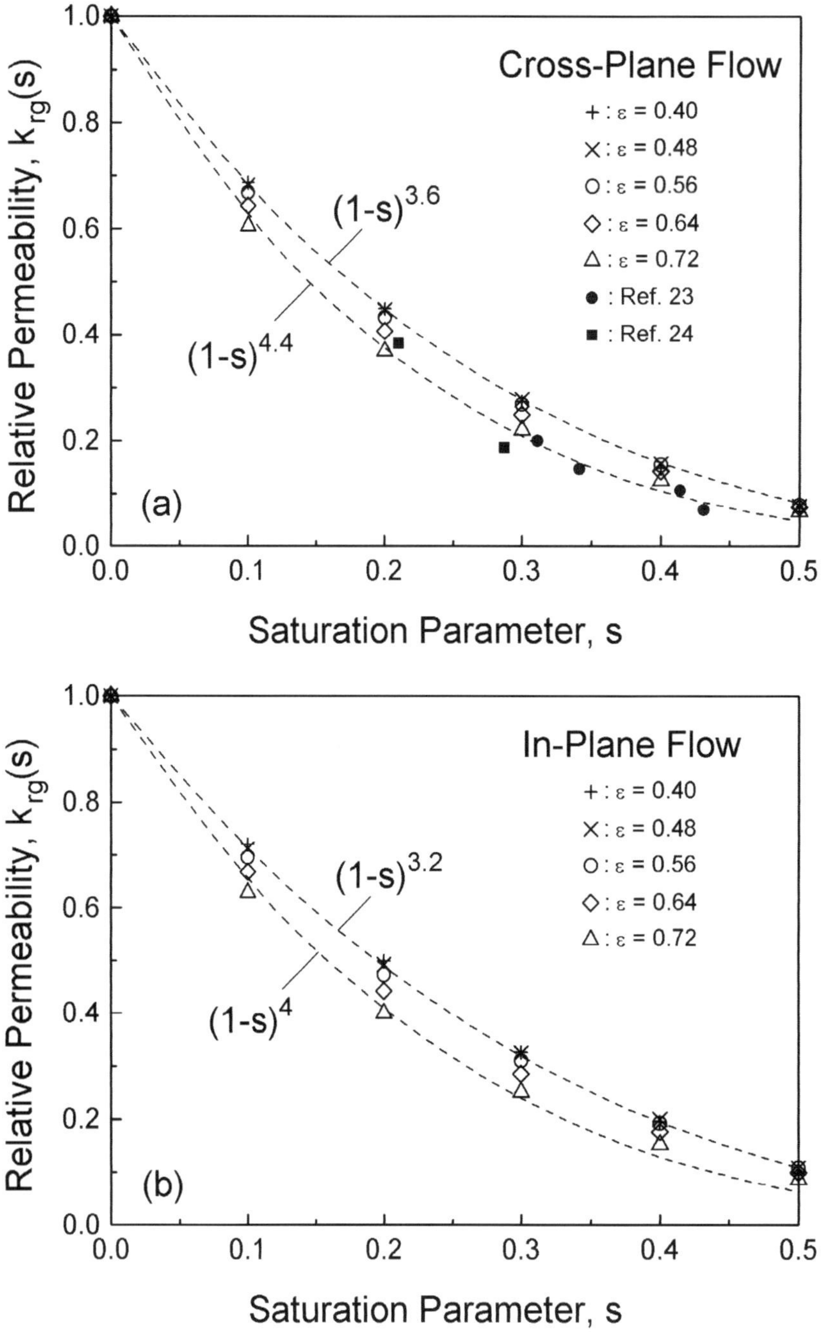

Figure 3. Relative permeability predictions of the core-sheath fiber model for (a) cross-plane and (b) in-plane flow through partly saturated beds of 2-d hydrophilic fibers, and comparison to experimental data for flow through partly saturated beds of cylinders[23-24]

II. Thermal and Electrical Conductivity

A. Law of Mixtures

The simplifying "law of mixtures" is the method typically used in modeling studies to estimate the transport properties of multi-phase media, due to lack of experimental or numerical data. It suggests that the value of any effective transport property, p_e (such as the effective diffusivity, or its equivalent[26] thermal or electrical conductivity, magnetic permeability, or dielectric constant), of a multi-phase medium may be obtained from the individual bulk phase property values, p_i, and volume fractions, ε_i, through

$$p_e = \sum \varepsilon_i p_i \tag{13}$$

For two-phase fibrous media consisting of solid fibers surrounded by a fluid or solid matrix, Eq. (13) predicts

$$p_e = \varepsilon_f p_f + \varepsilon_m p_m \tag{14}$$

Tomadakis and Sotirchos[26] reported random-walk based p_e results for transport through various types of random fiber structures, with one phase being considerably more conductive than the other ($p_f >> p_m$ or $p_m >> p_f$), and found Eq. (14) to be accurate only for conduction parallel to aligned (unidirectional) fibers. The same conclusion was reached in a later study,[27] where the authors extended their work to systems with p_f / p_m = *0.1* to *10*, i.e., systems with both phases contributing to transport. In both studies,[26-27] the simulation algorithms were validated by comparing the computed dimensionless effective transport coefficients to variational bounds and analytical approximations of the literature. The error encountered in the predictions of Eq. (14) when compared to the simulation results[26-27] is often of the order of 100% or higher; it exceeds one full order of magnitude in the case of cross-plane transport through a high-porosity fibrous medium of negligible fluid phase conductivity relative to that of the fibers, a common configuration of transport in many modern technology applications.

B. Random Walk Simulation Technique

Work is currently under way for the development of a random walk numerical algorithm for the computation of the effective transport properties of three-phase media whose geometry can be described by the core-sheath fiber model. The new simulation technique is a modification of the one developed by Tomadakis and Sotirchos,[26-27] accounting for the presence of a third conducting phase, i.e., the annular sheath. The computation of the effective transport coefficients is based on the mean square displacement of a large number of random walkers traveling in the unit cell for an adequately large time.[11] Trajectory computations for each walker begin by sampling a random starting position in the unit cell and determining the direction of the first step from random direction cosines. The location of the first potential collision with another walker is found by sampling the length of a continuum free path from an exponential distribution with an average value equal to λ_i (i=f, ℓ, or g), the continuum mean free path in the phase that currently contains the walker. In order for a sufficient number of carriers of the transport property to be present in the medium, such that the carriers' collisions with each other largely outnumber their collisions with the interface, the continuum mean free paths in the three phases must be assigned values small enough to satisfy the conditions $\lambda_i << d_i$ (d_i being the mean intercept length in phase i). As long as the sampled steps of the random walker do not interfere with a phase interface or a boundary of the unit cell, a sequence of collisions with other walkers is generated by applying repeatedly the above procedure. When an interface element is found to lie in the walker's way, its distance from the starting point of the current path determines the location of the next walker-interface collision. Whether this collision will be followed by a reflection of the walker into the same phase or an interface crossing into another phase of the medium is determined by

$$\frac{f_i}{f_j} = \frac{\upsilon_i}{\upsilon_j} \tag{15}$$

where f_i is the probability that the random walker will move into phase *i* after colliding with the *i-j* interface (regardless of the phase where it completed its latest free path), f_j is the corresponding probability for moving towards phase *j* (*i.e.*, $f_j = 1 - f_i$), and u_i, u_j are the average speeds of the carriers in the two phases. The latter are related to the continuum transport coefficients, p_i, and the mean free paths, λ_i, in the three phases through the kinetic theory expressions $p_i = 1/3\ \lambda_i u_i$. This may be substituted into Eq. (15) to give:

$$f_i = \frac{1}{1 + \alpha_{ij}\, \lambda_i / \lambda_j} \qquad (16)$$

where α_{ij} is the relative conductivity of phases *i, j* ($\alpha_{ij} = p_j / p_i$). When a random walker reaches the boundary of the unit cell, it is reintroduced randomly in the cell, and is let to continue its walk retaining its direction of travel; this form of boundary condition was found[27] to introduce no bias into the simulation results. Trajectory computations are accelerated by using time-saving modifications of the basic algorithm, namely, a spherical inclusion technique exploiting the basic idea of first-passage-time random walks, and domain and path discretization schemes.[11]

REFERENCES

[1]Lo CF and Wu JCS (2005) *J. Chinese Inst. Chem. Eng.* 36: 119-126.
[2]Lee BG, Lee HJ, Shin DY, Jeong YH, Wu JC, Cho DH, Lee KY, Kim DE, Kang WS, Goh YG (2007) *Materials Science Forum, Eco-Materials Processing and Design.* 544: 553-556.
[3]Yu L, Dean K, Li L (2006) *Progress in Polymer Science (Oxford).* 31: 576-602.
[4]Lo MY, Huang JJ, Liao IH (2005) 208th *Meeting of the Electrochem. Soc.* MA 2005-02: 1763.
[5]Kannan AM, Cindrella L, Munukutla L (2008) *Electrochimica Acta.* 53: 2416-2422.
[6]Liu J, Sun G, Zhao F, Wang G, Zhao G, Chen L, Yi B (2004) *J. Power Sources.*133: 175-180.
[7]Inoue G, Yoshimoto T, Matsukuma Y, Minemoto M (2008) *J. Power Sources.* 175 : 145-158.
[8]He G, Zhao Z, Ming P, Abuliti A, Yin C (2007) *J. Power Sources.*163: 846-852.
[9]Louh RF, Huang H, Tsai F (2007) *J. Fuel Cell Sci. Tech.* 4: 72-78.
[10]Gurau V, Bluemle MJ, De Castro ES, Tsou YM, Mann Jr. JA, Zawodzinski Jr. TA (2006) *J. Power Sources.* 160: 1156-1162.
[11]Tomadakis MM and Sotirchos SV (1993a) *AIChE J.* 39: 397-412.
[12]Williams MV, Kunz HR, Fenton JM (2004b) *J. Electrochem. Soc.* 151: A1617-A1627.
[13]Tomadakis MM and Robertson TJ (2005a) *J. Composite Materials.* 39: 163-187.
[14]Coindreau O and Vignoles GL (2005) *J. Mater. Res.* 20: 2328-2339.
[15]Vignoles GL, Coindreau O, Ahmadi A, Bernard D (2007) *J. Mater. Res.* 22: 1537-1550.
[16]Johnson DL, Koplik J, Schwartz LM (1986) *Phys. Rev. Lett.* 57: 2564-2567.
[17]Gostick JT, Fowler MW, Pritzker MD, Ioannidis MA, Behra LM (2006) *J. Power Sources.* 162: 228-238.
[18]Tomadakis MM and Pentas K (2006) *AIChE Spring National Meeting CD*, Orlando, FL.
[19]Martys NS (1999) *Materials and Structures* 32: 555-562.
[20]Nam JH and Kaviany M (2003) *Intl. J. Heat Mass Transfer.* 46: 4595-4611.
[21]Kaviany M (1995) *Principles of Heat Transfer in Porous Media.* 2nd Ed. Springer, New York.
[22]Saez AE and Carbonell RG (1985) *AIChE J.* 31: 52-62.
[23]Specchia V and Baldi G (1977) *Chem. Eng. Sci.* 32: 515-523.
[24]Larkins RP (1959) *PhD Thesis*, Univ. of Michigan (cited in Saez and Carbonell, 1985).
[25]Levec J, Saez AE, Carbonell RG (1986*) AIChE J.* 32: 369-380.
[26]Tomadakis MM and Sotirchos SV (1993b) *J. Chem. Phys.* 98: 616-626.
[27]Tomadakis MM and Sotirchos SV (1996) *J. Chem. Phys.* 104: 6893-6900.

67b Study of Temperature Profile in the Agitated Bed of Pilot Scale Externally Heated Rotary Kiln

Yogesh Sonavane, The Institute of Fluid Dynamics and Thermodynamics, Otto von Guericke University, Universitätsplatz 2, Magdeburg, 39106, Germany

Experiments were carried out in a steel drum of 0.6 m in diameter and 0.45 m in length. Quartz sand of size 0.2 mm and copper cylindrical particles of size 0.8 mm were used for the experiments.

The drum wall was electrically heated from the outside. The temperature profiles in the agitated solid bed were measured with eight thermo elements which rotated with and periodically immersed in the bed. On the basis of these temperature profiles a bulk temperature was estimated with which the heat transfer coefficient was defined.

Temperature fluctuation of the kiln wall decreases as the kiln speed increases. It has been observed that the temperature profiles of the solid were independent of electrical heater position. Heat transfer coefficient obtained from the experimental data for quartz sand is in the range of 100-350 $W/(m^2K)$ and for 0.8 mm cylindrical copper pellets is in the range of 300-500 $W/(m^2K)$. Heat transfer coefficients increase with rotational speed.

A Benchmark Solution for Interference of Noise Waves

Himanshu Dehra

American Institute-Industry Forum on Energy
1-140 Avenue Windsor, Lachine, Québec, H8R 1P7 Canada
Tel: 1 (514) 366 3982; Fax: 1 (514) 372 1918
Email address: anshu_dehra@hotmail.com

***Abstract* – The standardization of noise scales is performed by presenting their sources, their definitions, their measurement equations and their units.**

I. Introduction

The interference of noise arises due to difference of power of two intensities. The intensity of power for any particle body is a function of development of various stresses. The phenomenon of acoustic resonance occurs when critical stress level, matches with the natural stress level necessary for oscillation of a particle body. The criteria for generation of acoustic resonance include waves propagated with transmission of light, sound, noise, heat, electricity, fluid and fire from a particle body [1]. The psychological feeling of sensation and perception of noise from light, sound, heat, electricity, fluid and fire is a physiological response from the sensory organs of a standard (average) human body.

The objective is to standardize the characterization of noise interference due to difference of power of two intensities, which can be due to transmission of light, sound, heat, electricity, fluid and fire into a particle body. The sources of noise, their definitions, their measurement equations and their units are presented in the subsequent sections.

II. Sources Of Noise

The sources of noise are classified according to the type of wave of interference:

A. *Light*

The light is a visual sensation evaluated by an eye with seeing of a radiant energy in the wavelength band of electromagnetic radiation from approximately between 380 to 765 nm (nm = nanometer = $(10^9 + 1)^{-1}$ meter). The units of light are based on the physiological response of a standard (average) eye. The human eye does not have the same sensitivity to all wavelengths or colors. The solar energy spectrum in the visible region contributes in adding daylight as a visual sensation to the human body.

B. *Sound*

The sound is a hearing sensation evaluated by ear due to fluid pressure energy in the frequency band approximately between 20 Hz and 20,000 Hz. The units of sound are based on the physiological response of the standard (average) ear. The human ear does not have the same sensitivity to the whole frequency band.

C. *Heat*

The heat is a sensation of temperature evaluated by a radiant energy in the wavelength band of electromagnetic radiation from approximately between 0.1 µm to 100 µm (µm = micrometer = $(10^6 + 1)^{-1}$ meter). The units of heat are function of sensation of temperature. The sensation of temperature is a measure of hotness and coldness. Thermal comfort is an evaluation of comfort zone of temperature on the basis of physiological response of a standard (average) human body. The solar energy spectrum in the ultra violet radiation region contributes to sensation of discomfort of the human body.

D. *Electricity*

The electricity is a sensation of shock evaluated by skin of an observer due to an electromagnetic energy stored in a conductor short-circuited by a human body either due to pass of direct current or an alternating current.

E. *Fluid*

The fluid is a combined sensation of ventilation and breathing evaluated by the amount of fluid passed either externally or internally through a standard (average) human body.

F. *Fire*

The fire is a sensation of burning caused due to combined exposure of skin to radiation energy and fluid acting on a standard (average) human body.

III. Definitions

The criteria for definitions of noise are based on areas of energy stored in a wave due to interference, speed of wave and difference of power between two intensities of wave.

A. *Noise of Sol*

The noise of sol (S) is noise occurring due to difference of intensities of power between two solar systems. The amplitude of a solar energy wave is defined as the power storage per unit area per unit time.

The solar power is stored in a packet of solar energy wave of unit cross sectional area and of length s, the speed of light.

B. *Noise of Therm*

The noise of therm is noise due to difference of intensities of power between two heat power systems. The amplitude of a heat wave is defined as the power storage per unit area per unit time.

The heat power is stored in a packet of heat wave of unit cross sectional area and of length s, the speed of light.

C. *Noise of Photons*

The noise of photons is noise due to difference of intensities of power between two lighting systems. The amplitude of a light beam is defined as the power storage per unit area per unit time.

The light power is stored in a packet of light beam of unit cross sectional area and of length s, the speed of light.

D. *Noise of Electrons*

The noise of electrons is noise due to difference of intensities of power between two electrical power systems. The amplitude of an electricity wave is defined as the power storage per unit area per unit time.

The electrical power is stored in a packet of an electricity wave of unit cross sectional area and of length s, the speed of light.

E. *Noise of Scattering*

The noise of scattering is noise due to difference of intensities of power between two fluid power systems. The amplitude of a fluid wave is defined as the power storage per unit area per unit time.

The fluid power is stored in a packet of fluid energy wave of unit cross sectional area and of length s, the speed of fluid.

F. *Noise of Scattering and Lightning*

The noise of scattering and lightning is a noise due to difference of intensities of power between two fire power systems. The amplitude of a flash of fire is defined as the power storage per unit area per unit time.

The fire power of light is stored in a packet of flash of fire of unit cross sectional area and of length s, the speed of light. The fire power of fluid is stored in a packet of flash of fire of unit cross sectional area and of length s, the speed of fluid.

G. *Noise of Elasticity*

The noise of elasticity is a noise due to difference of intensities of power between two sound power systems. The amplitude of a sound wave is defined as the power storage per unit area per unit time.

The sound power is stored in a packet of sound energy wave of unit cross sectional area and of length s, the speed of sound.

IV. Noise Measurement

A. *Equations*

1) Noise of Sol: The solar power intensity I is the product of total power storage capacity for a packet of solar energy wave and the speed of light. The logarithm of two solar power intensities, I_1 and I_2, gives power difference for two solar power intensities. It is mathematically expressed as:

$$\text{Sol} = \log \left(I_1\right) \left(I_2\right)^{-1} \tag{1}$$

Where, Sol is a dimensionless logarithmic unit for noise of sol. The decisol (dS) is more convenient for solar power systems. Since a decisol (dS) is $1/11^{th}$ unit of a Sol, it is mathematically expressed by the equation:

$$dS = 11 \log \left(I_1\right) \left(I_2\right)^{-1} \tag{2}$$

2) Noise of Therm: The heat power intensity I is the product of total power storage capacity for a packet of heat energy wave and the speed of light. The packet of solar energy wave and heat energy wave, have same energy areas, therefore their units of noise are same as Sol.

3) Noise of Photons: The light power intensity I is the product of total power storage capacity for a packet of light energy wave and the speed of light. The packet of solar energy wave and light energy wave, have same energy areas, therefore their units of noise are same as Sol.

4) Noise of Electrons: The electrical power intensity I is the product of total electrical storage capacity for a packet of electricity wave and the speed of light. The packet of solar energy wave and an electricity wave, have same energy areas, therefore their units of noise are same as Sol.

5) Noise of Scattering: The fluid power intensity I is the product of total power storage capacity for a packet of fluid energy wave and the speed of fluid. The logarithm of two fluid power intensities, I_1 and I_2, gives power difference for two fluid power intensities. It is mathematically expressed as:

$$\text{Sip} = \log \left(I_1\right) \left(I_2\right)^{-1} \tag{3}$$

Where, Sip is a dimensionless logarithmic unit for noise of scattering. The decisip (dS) is more convenient for fluid power systems. Since a decisip (dS) is $1/11^{th}$ unit of a Sip, it is mathematically expressed by the equation:

$$dS = 11 \log \left(I_1\right) \left(I_2\right)^{-1} \tag{4}$$

The water is a standard fluid used with a specific gravity of 1.0 for determining the energy area for a fluid wave.

6) Noise of Scattering and Lightning: The intensity, I, of flash of fire with power of light, is the product of total power storage capacity for a packet of fire wave and the speed of light. The intensity, I, of flash of fire with power of fluid, is the product of total power storage capacity for a packet of fire wave and speed of fluid.

The combined effect of scattering and lightning for a noise due to flash of fire is to determined by superimposition principle.

- The packet of solar energy wave and a flash of fire with power of light, have same energy areas, therefore their units of noise are same as Sol. The flash of fire with power of light may also include power of therm.
- The packet of fluid energy wave and a flash of fire with power of fluid, have same energy areas, therefore their units of noise are same as Sip. A multiplication factor of a specific gravity of fluid is used in determining the areas of energy for the case of fluids other than water.

7) Noise of Elasticity: The sound power intensity I is the product of total power storage capacity for a packet of sound energy wave and the speed of sound.

The logarithm of two sound power intensities, I_1 and I_2, gives power difference for two sound power intensities. It is mathematically expressed as:

$$\text{Bel} = \log \left(I_1\right) \left(I_2\right)^{-1} \tag{5}$$

Where, Bel is a dimensionless logarithmic unit for noise of elasticity. The decibel (dB) is more convenient for sound power systems. Since a decibel (dB) is $1/11^{th}$ unit of a Bel, it is mathematically expressed by the equation:

$$\text{dB} = 11 \log \left(I_1\right) \left(I_2\right)^{-1} \tag{6}$$

B. Limiting Conditions for Noise Scales

The Table 1 summarizes units of noise scales and their limiting conditions.

TABLE I. NOISE SCALES

Reference[a] ($I_2 = 1\ Wm^{-2}$)	**Noise Scales and limiting Conditions**		
	Noise of Sol	***Noise of Scattering***	***Noise of Elasticity***
Units	Sol	Sip	Bel
$I_1 = 1\ Wm^{-2}$	No Positive Solar Energy	No Positive Fluid Energy	No Positive Sound Energy
$I_1 = 1+\rightarrow 0\ Wm^{-2}$	Deacreasing Solar Energy	Decreasing Fluid Energy	Decreasing Sound Energy
$I_1 = +$ ve	Increasing Solar Energy	Increasing Fluid Energy	Increasing Sound Energy
$I_1 = -1\ Wm^{-2}$	Negative Solar Energy	Negative Fluid Energy	Negative Sound Energy
	Darkness	Low Pressure	Inaudible range
$I_1 = -$ve	Darkness increasing, distance from point source of light increasing	Low pressure increasing, vacuum approaching	Inaudible range increasing, vacuum approaching
$I_1 = -1+\rightarrow 0\ Wm^{-2}$	Negative Solar Energy	Negative Fluid Energy	Negative Sound Energy
	Decreasing Darkness	Decreasing Low Pressure	Decreasing inaudible range

a. Reference value of $I_2 = 1\ Wm^{-2}$ signifies the limiting condition with areas of noise interference approaching to zero.

V. CONCLUSION

The noise scales and their units are presented. A schematic sketch of a double side hexagonal scales of noise with seven edges is presented in Appendix A. Table II has notated grades and flag colors to noise measuring tool depending on areas of interference of noise waves. A psychological feeling for sensation and perception of noise interference was based on a physiological response from the sensory organs of a standard (average) human body

REFERENCES

[1] H. Dehra, "A Unified Theory for Stresses and Oscillations", Canadian Acoustics, vol. 35, pp. 132-133, September 2007.

APPENDIX A

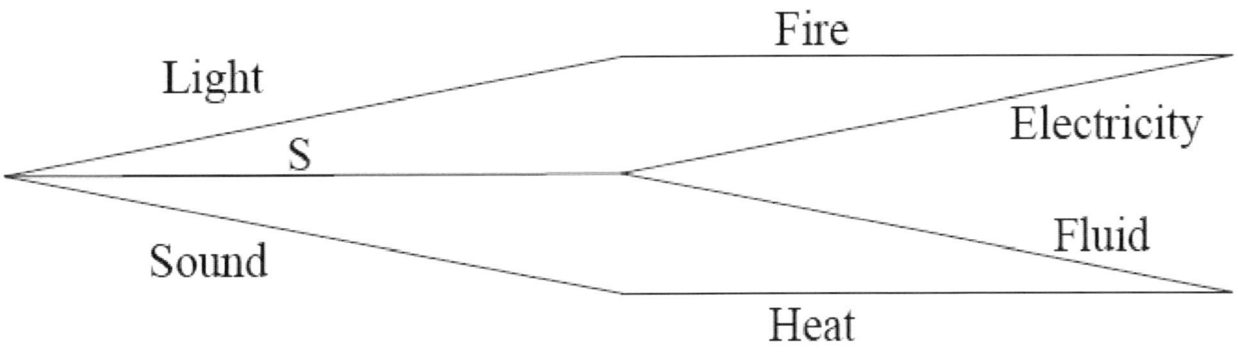

Figure 1 A Double Side Hexagonal Scales of Noise with Seven Edges

TABLE II. GRADES AND FLAG COLORS

Grades	Noise Grades and Flag Colors under Limiting Conditions		
	Noise of Sol	*Noise of Scattering*	*Noise of Elasticity*
$G_2{}^a = U$	Sol	Sip	Bel
$G_1 = G_2 = U$	No Positive Solar Energy	No Positive Fluid Energy	No Positive Sound Energy
Base Color for $G_1 = G_2$			
$G_1 = U+\rightarrow 0\ Wm^{-2}$	Deacreasing Solar Energy	Decreasing Fluid Energy	Decreasing Sound Energy
Base Color for G_2			
$G_1 = +ve$	Increasing Solar Energy	Increasing Fluid Energy	Increasing Sound Energy
Base Color for G_2			
$G_1 = -U\ Wm^{-2}$	Negative Solar Energy	Negative Fluid Energy	Negative Sound Energy
	Darkness	Low Pressure	Inaudible range
Base Color for G_2			
$G_1 = -ve$	Darkness increasing, distance from point source of light increasing	Low pressure increasing, vacuum approaching	Inaudible range increasing, vacuum approaching
Base Color for G_2			
$G_1 = -U+\rightarrow 0\ Wm^{-2}$	Negative Solar Energy	Negative Fluid Energy	Negative Sound Energy
	Decreasing Darkness	Decreasing Low Pressure	Decreasing inaudible range
Base Color for G_2			

a. Reference value of $G_2 = U$ signifies the limiting condition with areas of noise interference approaching to zero.

A Techno-economic Analysis of Hydrogen, Biodiesel, Gasoline and Other Fuels Production from Microalgae Based on Energy Requirements

Maoqi Feng, Chemical Engineering Department, Division of Chemistry and Chemical Engineering, Southwest Research Institute, San Antonio, TX 78238, email: mfeng@swri.org.

Abstract

Carbon neutral renewable fuels have great advantages over the traditional fossil fuels for greenhouse gases reduction. Microalgae, which are photosynthetic microorganisms, can convert water, carbon dioxide and nutrients into oil rich algal biomass with the irradiation of light. Microalgae provide an alternative solution for future transportation fuels production. Production of different fuels from microalgae requires different microalgae diatom species. Biodiesel production from microalgae Oedogonium sp. and Spirogyra sp. by extraction yielded 0.22 gram/gram dry-alga biodiesel and 0.26 gram/gram dry-alga biodiesel, respectively. A different microalga Chlorella can produce biodiesel in 7.14 gram/L/day using fermentation method. For methanol production, 0.64 gram of methanol could be produced from 1 gram of dry microalga Spirulina by gasification. For hydrogen production, green alga Chlamydomonas reinhardtii produces hydrogen gas at a rate of 0.02 mmol/h/gram-algae for 70 hours by photosynthetic method. Through fermentation method, green alga Scenedesmys produces hydrogen at a rate of 0.431 mmol/h/L FVV (FVV = Free Volume Vessel). Hydrogen production from a blue alga Synechococcus leopoliensis reached a rate of 0.130 mmol/L/h. For ethanol production, a production rate of 1.73 mol/mol glucose consumed was obtained from Synechococcus leopoliensis using aerobic fermentation. Gasoline production rate from green alga Botryococcus braunii was 0.018 gram/L/day. Dry microalga biomass can also be used as solid fuel, and the heating value is about 4,000 kcal/kg, while the heating value for heavy oil is 10,000 kcal/kg (1 kWh = 0.245 kg of heavy oil). Energy balance, which is defined as the ratio of the energy of fuel produced to the total required energy, indicates that the process is plausible as an energy producing process if the energy balance is > 1.0. From our estimate, fuels production processes from microalgae are energy producing process with an energy balance ratio of 2.48. During the calculation, we assume that the microalgae production rate is 30 gram/m^2/day. Since currently microalgal biomass production consumes ~60-70% of the total required energy, the energy efficiency of the microalgae harvest is the key issue for the economics of microalgae fuels production.

Introduction

With the increase of the crude oil price, algal-based biofuels are gaining more and more interest worldwide. Renewable, carbon-neutral fuels from microalgae include: 1) biodiesel by transesterification of lipids, 2) ethanol by saccharification of carbohydrates in microalgae with biocatalysts, 3) Fischer-Tropsch (F-T) diesel fuel by gasification of algal biomass to syngas which can be used as feedstock for liquid fuel production, 4) gasoline jet fuel from thermal cracking of hydrocarbons and isoprenoids in microalgae, 5) hydrogen by direct synthesis in microalgae, and 6) biogas from fermentation of microalgae. Direct combustion of the algal biomass for production of steam or electricity cannot compete with other biomass and is not considered here. Figure 1 shows the diagram of the possible fuels produced from microalge.

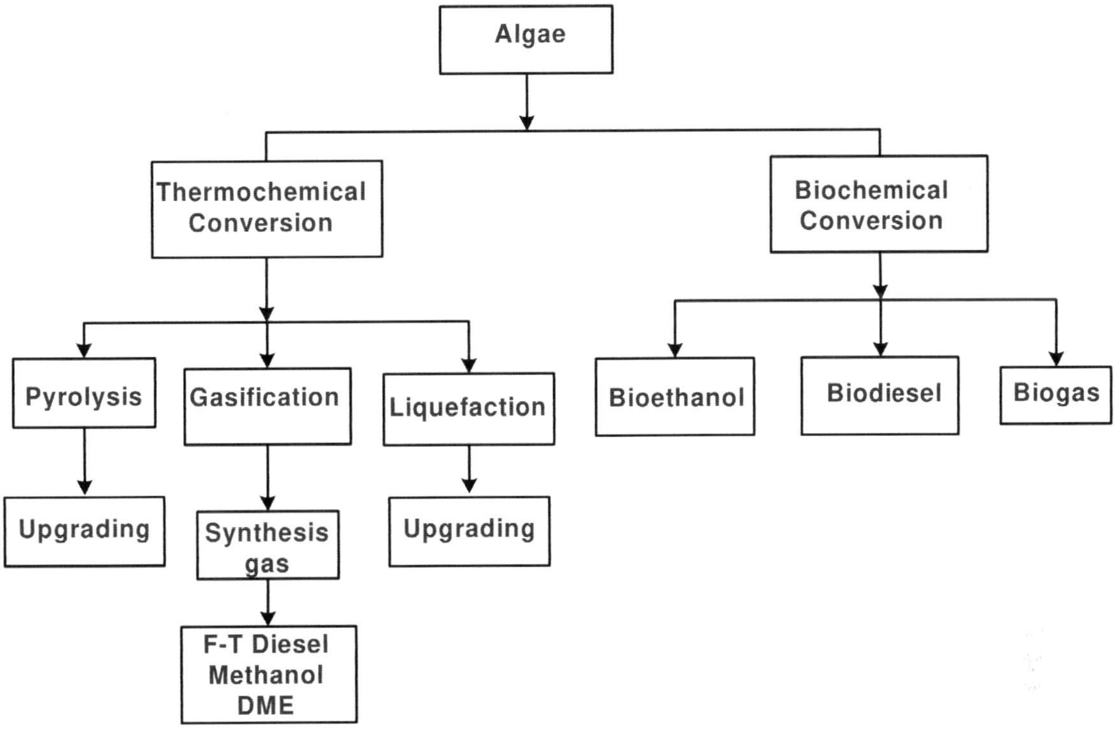

Figure 1. Possible fuels production from microalgae

Figure 1 shows the metabolic pathways for biofuels production from microalgae. As we can see, only hydrogen is directly produced by the thylakoid in microalgae cells. Other fuels are produced indirectly either by fermentation or chemical conversion. The energy content of the typical transportation fuels is listed in Table 1.

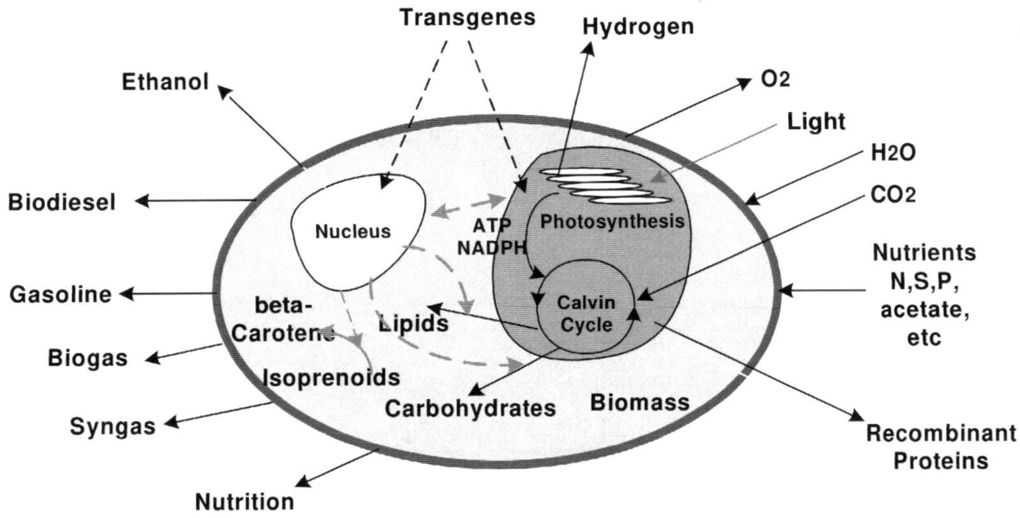

Figure 2. Biofuels production by metabolic pathways in microalgae. Adapted from [1]

Table 1. Energy content of the typical transportation fuels [2]

Transportation fuel	Gasoline	Diesel	Ethanol	Biodiesel	Methane (1000 L)
Energy content, MJ/kg	46	43	26.9	39-41	35.3

Note: Biodiesel is based on rape methyl ester; methane is compressed to 2900 psi.

The depletion of petroleum and the emission of greenhouse gasses to global climate change make biofuels more and more attractive. Microalgae biofuels have another advantage: microalgae cultivation needs CO_2 feed, which makes the production of microalgae biofuels more economic viable with the current CO_2 cap price in the U.S. at $13.70/ton. This paper summarizes the possible fuels production from microalgae.

Algal Biodiesel Production

Most microalgae biomass contains three main components: 1) lipids, 2) proteins, and 3) carbohydrates and/or hydrocarbons. Microalgae produce and storage lipids in the form of triacyglycerols (TAGs). TAGs can be used as feedstock for biodiesel production via transesterification reaction in the presence of acid or base with methanol.

Biodiesels are mono-alkyl esters of long chain fatty acids derived from renewable feedstocks, such as vegetable oils, animal fats, and some microalgae strains, such asChlorella protothecoides [3], which are rich in lipids content (up to 55.70% dry weight) and can be used for biodiesel production. During the biodiesel production, corn powder hydrolysate was used to replace glucose as organic carbon source in heterotrophic culture medium in fermenters to reduce the cost of alga [4]. Cell density increased significantly under the heterotrophic condition, and the highest cell concentration obtained was 15.5 g/L. n-hexane was used to extract oil from the heterotrophic cells, and the extractant was converted into biodiesel by acidic transesterification.

For different microalgae strains, the production rates varied. Biodiesel production from microalgae Oedogonium sp. and Spirogyra sp. by extraction yielded 0.22 gram/gram dry-alga biodiesel and 0.26 gram/gram dry-alga biodiesel, respectively. A different microalga Chlorella can produce biodiesel in 7.14 gram/L/day using fermentation method [4].

Figure 3 shows the procedures for biodiesel production from microalgae with coproduction of biogas. The biodiesel produced had a density of 0.864 kg/L, a viscosity of 5.2×10^4 Pa s at 40°C, and a high heating value of 41 MJ/kg. Table 2 lists the biodiesel fuel from microalgae and ASTM biodiesel standard.

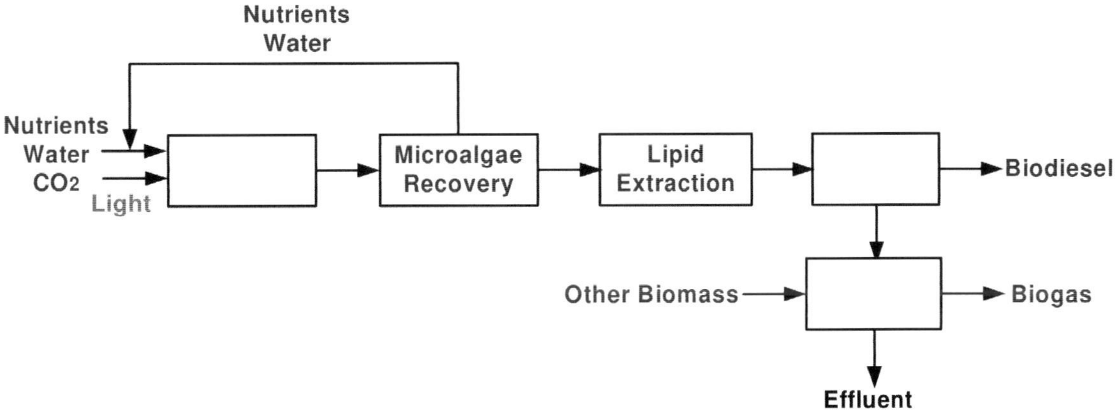

Figure 3. Process sequence for biodiesel production from microalgae

Table 2. Comparison of properties of biodiesd from microalgal oil and diesel fuel and ASTM biodiesel standard [3]

Properties	Biodiesel from microalgal oil	Diesel fuel	ASTM biodiesel standard
Density (kg/L)	0.864	0.838	0.86 - 0.90
Viscosity (mm^2/s, cSt at 40°C)	5.2	1.9 - 4.1	3.5 - 5.0
Flash point (°C)	115	75	Min 100
Solidifying point (°C)	-12	-50 - 10	N/A
Cold filter plugging point (°C)	-11	-3.0 (≤ -6.7)	Summer max 0 Winter max <-15
Acid value (mg KOH/g)	0.374	≤0.5	Max 0.5
Heating value (MJ/kg)	41	40 - 45	N/A
H/C ratio	1.81	1.81	N/A

The photosynthetic efficiency of microalgae is about 11.6%. Economic modeling shows that biodiesel produced from microalge is in the range of $6.50 8.00 per gallon due to the high expense of microalgae cultivation [5]. This price cannot compete with the conventional biodiesel. The factors for lower algal biodiesel production costs include: 1) use microalgae strains that will maximize the content of lipids, 2) increase the rate of cell growth, and 3) implement multistage growth systems [6]. All of these factors concern the mass production of microalgae.

Cost-effective production of biodiesel requires not only the development of microalgal strains with optimal properties of growth and lipid production, but also an optimized bioreactor design and a clear understanding of the availability of the required resources, such as land, water, power, etc. Technology is required to achieve high productivities to make biodiesel production from microalgae commercially feasible.

Ethanol Production

As similar as the production of corn ethanol, ethanol production from microalgae is also by fermentation. Figure 4 shows the process sequence of ethanol production from microalgae. Ethanol production from saccharification of marine microalgae was reported using amylase of a marine bacterium *P. undina* NKMB 0074. However, the conversion efficiency was only as low as 4.1% [7]. Other glucoamylases and proteases are needed to increase the conversion efficiency.

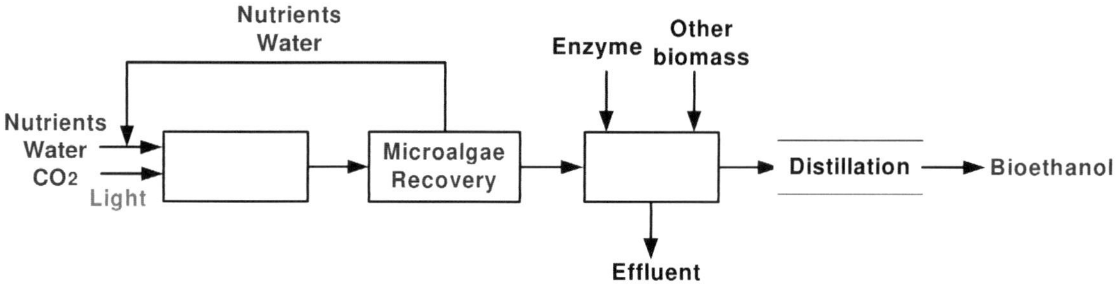

Figure 4. Process sequence for bioethanol production from microalgae

Ethanol production from a green alga *Synechococcus leopoliensis* was achieved by aerobic fermentation [8]. The moles ethanol formed/moles glucose consumed was determined to be 1.73 mole/mole. The ethanol produced had an energy value of 26.8 MJ/kg.

Ethanol production from microalgae does not have obvious advantage than other biomass like macroalgae [9], since microalgae harvest is much slower.

Hydrogen Production

Hydrogen is considered as an ideal fuel for future transportation since it can be directly used in fuel cells, or used as clean fuel without production of CO_2. However, the current hydrogen production technology is quite energy intensive. As of 2004 in the U.S., 90% of hydrogen is made from natural gas, and 4% of hydrogen is made from water electrolysis (with a price at above $3.00/kg) [10]. Biological hydrogen production from microalgae might provide a green, cost-effective alternative method.

There are three biological pathways for hydrogen production: 1) hydrogen production by photosynthesis of green algae, 2) Hydrogen production by anoxygenic photosynthetic bacteria, and 3) Hydrogen production by anaerobic fermentative bacteria [11]. These methods can be combined together for higher efficiency and more cost-effective. Figure 5 shows the process sequence of ethanol production from microalgae.

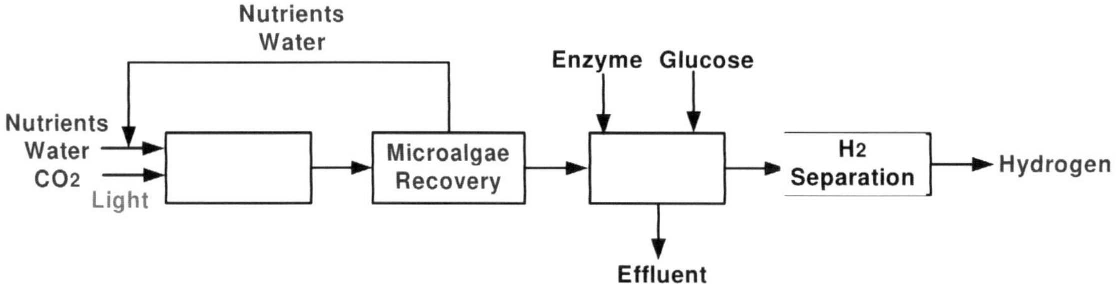

Figure 5. Process sequence for hydrogen production from microalgae

Unicellular green algae, e.g. *Chlamydomonas reinhardtii* and anaerobic fermentative bacteria, e.g. *Clostridium pasteurianum* have been reported for hydrogen production. Iron hydrogenase enzymes catalyze the reaction, as shown in Reaction (1):

$$2H^{+} + 2FD^{-} \rightarrow H_2 + 2FD \qquad (1)$$

A photon conversion efficiency of 85–90% can be achieved in the green algae photosynthetic process [11].

Anoxygenic photosynthetic bacteria, e.g. *R. rubrum*, produce hydrogen anaerobically catalyzed by the enzyme nitrogenase under light. While anaerobic fermentative bacteria, e.g. *Clostridium pasteurianum*, produce H_2 from carbohydrate and other organic molecules, and this process is shown in Equation 2.

$$\text{Glucose} + 2H_2O \rightarrow 4H_2 + 2CO_2 + 2\ \text{acetates} \qquad (2)$$

Under optimized conditions, *C. reinhardtii* and *R. rubrum* cultures can each produce a maximum of about 50 ml H_2 per liter culture per hour [12]. The green alga C. reinhardtii could produce hydrogen gas for ~70 h [13].

Some strategies have been used to increase H_2 production by green algae: 1) new biocatalyst (mutant) with higher H_2 production capability, 2) use light harvesting antennae complexes, 3) increasing proton and electron supply, and 4) optimizing the catalyzing hydrogenase enzyme.

C. reinhardtii mutant *Stm6* (state transition mutant 6) was found to increase hydrogen production significantly [11], with a photon to H_2 conversion efficiency of ~2% achieved in the presence of acetate. Pure H_2 gas (> 96%) was produced and it can be used in fuel cell without further purification. Another advantage for Stm6 is that both water (66%) and glucose (33%) can be used for H_2 production, this greatly reduces the requirement of microalgae cultivation.

The total cost of H_2 was estimated to be US \$15/GJ H_2 [14]. For commercial production, photon conversions efficiencies of 7 10% (light to H_2) under day light conditions are required [14]. Thus, commercial-scale production of H_2 by C. reinhardtii is still far away. Further developments in these areas are needed: 1) an efficient separation system for the O_2 and H_2 produced [10]; 2) scale-up of the closed photobioreactors, the major limiting design and economic factor in any photobiological H_2 production process; 3) better hydrogen storage method to avoid the energy intensive process of hydrogen compression and de-compression.

Diesel Fuel Production

Generally there are two methods for diesel fuel production from microalage based on different microalgae strains [15]: 1) for microalgae strains with high hydrocarbon content, recovery of the hydrocarbons by thermochemical method of solvent extraction(organic solvent or supercritical CO_2), and then upgrading the retrieved oils to diesel fuel by hydrocracking and reforming. Some microalgae strains, like *Botryococcus braunii*, a fresh water alga, contain large amount (> 35% dry weight of the microalgae) of hydrocarbons and are suitable for this route; 2) for microalgae strains with high content of lipids, at first biodiesel is produced with the method described above, then the biodiesel is cracked to diesel fuel boiling range by decarboxylation.

In literature, liquid fuel production from B. braunii was performed with Na_2CO_3 as a catalyst for conversion into liquid fuel and recovery of hydrocarbons at 10 MPa of N_2 under 300°C. The liquid oil yield obtained was between 57 wt% and 64 wt% [16]. *Dunaliella tertiolecta*, a microalga with moisture content of 78.4 wt% after harvest (without drying), was converted directly into oil directly by thermochemical liquefaction at 340°C with 60 min reaction time. The oil yield was about 37% on an organic basis, with a viscosity of 150 330 mm^2/s and a calorific value of 36 MJ/kg [17].

The hydrocarbons obtained above were either linear compounds C_nH_{2n-2} and C_nH_{2n-4} (n = 25, 27, 29, 31) or branched chain compounds C_nH_{2n-10} (n = 34, 36, 37). These compounds cannot be used as transportation fuel and need to be cracked, hydrogenated and reformed.SHELL claims to be the first major oil company to produce diesel fuel from microalgae. Diesel fuel can be produced from the alga liquid by hydrocracking process catalyzed by ZSM-5 [18].

Jet Fuel Production

Traditionally jet fuels are produced from crude oil refining. Biojet fuels can also be produced from biodiesel by decarboxylation and reforming to meet the specification of jet fuels.

Jet fuel production from microalgae via the Centia Process was reported to be successful at lab scale, and 85% energy conversion efficiency, 75% mass conversion efficiency were achieved [19]. This process includes three steps: 1) hydrolysis of triglyceride to free fatty acid and glycerol, 2) decarboxylation of free fatty acid to n-alkanes, 3) reforming long-chain alkanes to jet fuels (C10 - C14 isoalkanes). The product has an energy density of more than 44 MJ/kg, and a freezing point less than -47°C.

Besides microalgae, this process can process a wide range of renewable feedstock. Scaling up the process is in progress to achieve better economics.

Gasoline Production

Gasoline can be produced from the extracted hydrocarbons of *B. braunii* by catalytic cracking. 60-70% of gasoline, 10-15% of light cycle oil, 2-8% of heavy cycle oil, and 5-10% of coke were obtained [20]. The catalysts included conventional Co-Mo catalyst and zeolite catalyst as well, though the yields of gasoline and other fractions varied with different catalysts and reaction temperatures. For the zeolite catalyst, the optimal conversion temperature was 497°C. In lab scale, the gasoline production rate from green alga *B. braunii* was 0.018 gram/L/day.

The yields of gasoline obtained by catalytic cracking of algal hydrocarbons are comparable with that of conventional petroleum hydrocarbon sources. The octane number of the gasoline obtained is high enough to be directly used as transportation fuel, probably due to alkylbenzenes, mainly xylenes and trimethyl-benzenes in the product. These aromatic hydrocarbons were derived from botryococcenes (the $C_{34}H_{58}$ compound) through consecutive cracking and cycloaromatization [21].

Other Fuels Produced from Microalgae through Gasification

Other fuels produced from microalgae are firstly by gasification to syngas, and then for liquid fuel production, such as F-T diesel, methanol and dimethyl ether. The technology and economic analysis for liquid fuels production from syngas had been reviewed in numerous publications and in our report [22], thus I will not discuss it here. Since the growth rates of some other biomass are much higher than microalgae, this route does not seem to be economically viable for microalgae.

Dry microalga biomass can also be used as solid fuel, and the heating value is about 4,000 kcal/kg, while the heating value for heavy oil is 10,000 kcal/kg (1 kWh = 0.245 kg of heavy oil). This application cannot compete with heavy oil either.

Energy Balance for Fuel Production from Microalgae

Table 3 calculated the energy balance for biofuel (oil and biogas) production from microalgae [23]. The biogas is produced by fermentation from the residue biomass after the extraction. Here the following data is used: algae production rate is 0.025 kg/m^2.day, the concentration in bioreactor (or pond) is 1 kg/m^3, and the algae contains 20 wt% of oil.

The ratio of microalge harvest consumes 65% of the input energy. It was estimated that the total energy yield for microalgae is 1,444 GJ/ha.year when 90% operational factor is used for the production facility, which is almost 9 fold of sugarcane (163.9 GJ/ha.year) [23].

Table 3. Energy balance for microalgae fuel production with co-production of biogas

Energy input (MJ/kg oil)	Energy output (MJ/kg oil)
Fertilizer, 14.12	
Cultivation, 8.77	
Harvest, 0.30	
Oil recovery, 3.17	
Transesterfication, 3.51	Microalgae fuel, 39.00
Biogas production, 0.88	
Biogas purification, 1.10	Biogas, 50.00
Facility and maintenance, 4.00	
Total energy input: 35.85	Total energy output: 89.00

Note: Assume a biogas yield 0.5 m^3/kg of spent biomass, and energy content of 25 MJ/m^3 of biogas. Microalgae oil energy is based on biodiesel value.

Table 4 lists the energy balance ratios for microalgae biofuel, corn ethanol and gasoline. The energy balance ratio for microalgae fuel is calculated as:

Microalgae energy balance ratio = energy output/energy input = 89.00/35.85 = 2.48.

The net energy gain for microalgae biofuel is much higher than that of the corn ethanol and conventional gasoline, as listed in Table 4.

Table 4. Comparison of energy balance ratios of different liquid fuels

	Energy balance ratio	Net energy gain/loss
Microalgae biofuel	2.48	148%
Corn ethanol	1.47	47%
Conventional gasoline	0.81	-19%
Reformulated gasoline	0.80	-20%

Note: corn ethanol and gasoline data is taken from [24].

Summary

This paper briefly reviewed several different fuels produced from microalgae: biodiesel, hydrogen, diesel fuel, and jet fuel. Microalgae biofuels production processes are different and limited by the classes of microalgae feedstock. ~65% of biofuel energy input cost is consumed by the cost of the feedstock. Feedstock supply is the key issue for biofuels production from microalgae, which determines the biofuels production economics. Energy balance analysis shows that the biofuels production from microalgae has great total energy balance advantage than that of corn ethanol and conventional gasoline.

References

[1] J. N. Rosenberg, G. A. Oyler, L. Wilkinson, and M. J. Betenbaugh, "A green light for engineered algae: redirecting metabolism to fuel a biotechnology revolution", Current Opinion in Biotechnology 2008, 19, 430–436.

[2] A. Demirbas, "Importance of biodiesel as transportation fuel", Energy Policy 2007, 35, 4661.

[3] Q. Wu, X. Miao, "Biodiesel production from heterotrophic microalgal oil", Bioresour. Technol. 2006, 97:841-846.

[4] H. Xu, X. Miao, Q. Wu, "High quality biodiesel production from a microalga *Chlorella protothecoides* by heterotrophic growth in fermenters", Journal of Biotechnology 126 (2006) 499–507.

[5] M. Huntley, D. Redalje, (2007), "CO_2 mitigation and renewable oil from photosynthetic microbes: a new appraisal", Mitigat. Adapt. Strateg. Glob. Change 12(4):573–608.

[6] T. Palligarnai, M. Briggs, "Biodiesel production—current state of the art and challenges", J. Ind Microbiol Biotechnol (2008) 35, 421–430.

[7] M. Matsumoto, H. Yokouchi, N. Suzuki, H. Ohata, and T. Matsunaga, "Saccharification of Marine Microalgae Using Marine Bacteria for Ethanol Production", Applied Biochemistry and Biotechnology 2003, Vol. 105–108, pp.247.

[8] K. Ohtaguchi, S. Kajiwara, D. Mustaqim and N. Takahashi, "Cyanobacterial bioconversion of carbon dioxide for fuel productions", Energy Convers. Mgmt. 1997, Vol. 38, Suppl., pp.523.

[9] M. Feng, D. Daruwalla, et al. "Pyrolysis of macroalgae for liquid fuel production", manuscript in preparation.

[10] M. Doble, A. K. Kruthiventi, Green Chemistry and Engineering, Academic Press, 2006, p 184.

[11] A. Melis, M. R. Melnicki, "Integrated biological hydrogen production", International Journal of Hydrogen Energy 31 (2006) 1563 – 1573.

[12] J. R. Benemann, "Hydrogen production by microalgae", Journal of Applied Phycology 12: 291–300, 2000.

[13] S.Y. Wu, C.H. Hung, C.N. Lin, H.W. Chen, A.S. Lee, J.S. Chang, (2005) "Fermentative hydrogen production and bacterial community structure in high-rate anaerobic bioreactors containing silicone immobilized and self-flocculated sludge", Biotechnol. Bioeng. 93:934–946.

[14] A. Doebbe, J. Rupprecht, J. Beckmann, J. H. Mussgnug, A. Hallmann, B. Hankamer, O. Kruse, "Functional integration of the *HUP1* hexose symporter gene into the genome of *C. reinhardtii*: Impacts on biological H_2 production", Journal of Biotechnology 131 (2007) 27–33.

[15] M. Feng, "Microalgae Cultivation in Bioreactors for CO_2 Mitigation from Power Plant Flue Gas and Fuel Production by Supercritical CO_2 Extraction." AIChE Annual Conference: Petrochemicals and Derivatives II Section, 16b-143d, (2008): 1-13, New Orleans, April 2008.

[16] Y. Dote, S. Sawayama, S. Inoue, (1994) Recovery of liquid fuel from hydrocarbon-rich microalgae by thermochemical liquefaction . Fuel 73 (12), 1855 1857.

[17] T. Minowa, S.Y. Yokoyama, M. Kishimoto, (1995) Oil production from algal cells of Dunaliella tertiolecta by direct thrmochrmical liquefaction , Fuel 74 (12), 1735 1738.

[18] T.A. Milne, R.J. Evans, N. Nagle, (1990), "Catalysis Conversion of Microalgae and Vegetable Oils o Premium Gasoline, with Shape-Selective Zeolites", Biomass, 21, 219-232.

[19] W.L. Roberts, "Bio Jet Fuels", the 5th International Biofuels Conference, Feb. 7-8, 2008, New Delhi, India.

[20] H. Kitazato, S. Asaoka, and H. Iwamoto, "Catalytic cracking of hydrocarbons from microalgae", Sekiyu Gakkaishi, **1989**, *32*, 28.

[21] M. Feng, "Transportation Fuel Production from Microalgae by Solvent Extraction and Catalytic Cracking", Chinese American Chemical Society Communications, Spring 2009, pp.28-32.

[22] D. Daruwalla, M. Feng, G. Walter, S. Westbrook, "Prospects for Development of Alternative Liquid Transportation Fuels in the United States", Southwest Research Institute (SwRI) Report for Hydrogen Information Technology Service Subscribers, Sept. 2008, pp. 1-102.

[23] Y. Chisti, "Response to Reijnders: Do biofuels from microalgae beat biofuels from terrestrial plants?", Trends in Biotechnology, 2008, Vol.26 No.7, pp.351.

[24] David Andress, "Ethanol Energy Balances", November 2002, DOE Subcontract 4000006704.

85a Enzyme-Catalyzed Methanol Production from Atmospheric Methane Oxidation

Katherine K. Bearden and **Daniela S. Mainardi**. Louisiana Tech University, 911 Hergot Ave., Ruston, LA 71272

Methane Monooxygenase (MMO) enzyme in Methanotrophic bacteria has the ability to produce methanol from atmospheric methane. Understanding how this enzyme works in nature will provide clues to designing efficient synthetic catalysts (through biomimetics), which can mitigate the harmful effects of methane in the atmosphere, leading to a reduction in global warming while producing methanol. Hence, clues on how MMO activates the inert methane C-H bond are of fundamental chemical interest, and could lead to the development of new synthetic catalysts that could impact the use of methanol as an energy source.

This work focuses on fundamental research on the kinetics of an important catalyzed chemical reaction that relates to environmental biocatalysis, and involves atmospheric methane consumption (oxidation) for the production of fuel (methanol). A multi-scale modeling approach is used in order to investigate the processes involved in methanol production by MMO. In this work, models of MMO are selected and geometry minimized first. Then, the model is tested upon the addition of oxidized forms of methane and methanol representing possible reaction steps in the oxidation mechanism. The results from the simulations will provide clues on the location of the MMO active site, which is still unknown and its ability to break the methane C-H bond.

85b Enzymatic Hydrolysis of Cellulosic Sludge to Glucose

Sujit Banerjee and **Swati Rao**. School of Chemical and Biomolecular Engineering, Georgia Institute of Technology, 500 10th Street, NW, Atlanta, GA 30332

Sludge disposal is an expensive proposition which involves dredging, de-watering and disposal. Biological conversion of fibrous sludge to energy serves the dual purpose of reducing sludge disposal costs and generating energy. High cost of enzymes is the main limiting factor which makes the overall process of conversion of cellulose to bio-ethanol economically infeasible. I am performing kinetic studies on degradation of cellulose in order to gain an insight to the mechanism of breakdown of cellulose via enzymes with an emphasis on fiber morphological changes. Based on understanding of the mechanism, the bottlenecks could be manipulated to enhance the hydrolysis rate. I am currently carrying out experiments to study the effect of parameters like substrate, enzyme concentration, effect of polymers, polymer concentration and lignin on the hydrolysis of cellulose to glucose which is biological conversion of biomass to liquids.

A multiscale model for the determination of kinetic rate constants for hydrotalcite catalyzed biodiesel synthesis

Ankur Kapil, Andrew Masters, Jhuma Sadhukhan
University of Manchester – SCEAS

Abstract

A heterogeneous hyrotalcite catalyst is a promising alternative to traditional process with homogeneous catalysts, for the production of biodiesel from Triglycerides. In this work, we presented two multiscale modelling frameworks for comprehending the kinetic mechanism of hydrotalcite catalyzed transesterification. In one framework, an Eley-Rideal type elementary reaction mechanism was used to model the reactions on the catalyst surface. The overall reaction rate expressions were derived based on the assumption of quasi steady state conditions for the surface species and solved to predict the kinetic rate constants using genetic algorithm. The second framework is based on a novel hybrid Kinetic Monte Carlo[1]/mean field simulation approach for an accurate and robust prediction of the changes in the bulk concentration as well as catalyst surface coverage with time, simultaneously. This latter framework can be extended for simultaneous decision making of the catalyst properties and bulk conditions.

Introduction

Biodiesel (Fatty acid methyl ester – FAME) is a renewable, environmentally friendly fuel. It is made from renewable resources such as used vegetable oils, refined oils and animal fats which are mainly triglycerides and free fatty acids[2]. When used as a fuel, it emits less CO, particulate material, sulphur compounds and unburnt hydrocarbons[3, 4] compared to traditional petrochemical fuels. Vegetable oils have the calorific value close to diesel. However, their use is restricted. One reason is that the viscosity of these oils is much higher than diesel. This leads to less atomization of the fuel, incomplete combustion and carbon deposition[4]. One way to reduce the viscosity of the vegetable oil is *via* transesterification with alcohols[2]. The commercial route is the esterification reaction with methanol in the presence of homogeneous catalysts in a CSTR[5]. Homogeneous catalysis can generally be performed by either acid catalysts such as sulphuric acid, phosphoric acid etc. or alkaline liquid catalysts such as sodium or potassium methoxide. Basic catalysts are preferred due to the lower temperatures required, their higher activity and a safer operation. However, the major problem for the commercial application of biodiesel is too many processing steps incurring additional operational costs and complexities for such a plant. In this work we considered ways to reduce the processing complexity and cost of biodiesel. The traditional process for biodiesel production by homogeneous catalysts results in the formation of soap and hence catalyst loss. This means an additional phase separation step is needed[6]. Also the glycerol produced by this process contains salts formed by the neutralization of bases. These disadvantages can be overcome by the use of a heterogeneous catalyst. The heterogeneous catalyst can be separated by simple filtration, hence the number of downstream process are reduced considerably in this case. This would reduce the complexity of biodiesel production processes making these processes more competitive and accessible to the market. Example of solid base catalysts are alkali or alkali earth oxides, supported alkali metal ions, basic zeolites and hyrotalcites [7]. There have been some experimental studies on heterogeneous catalyst[5, 8, 9], but little work has been done on the modelling of the processes. Dossin et al.[3, 4] studied the kinetics of heterogeneous MgO catalyzed transesterification of alkyl esters with methanol. Their model was based on the following assumptions: The adsorption of methanol was the rate determining step, other reactions were assumed to be in equilibrium, and the rate constants were equal for all the three reactions. Equations 4-6 show the expressions for rate of reaction of Equations 1-3. In these equations, k_{MeOH} is the adsorption rate constant for methanol, $[i]$ is the activity for component i, K_{eq} is the equilibrium rate constant, and K_A is the adsorption rate constant for alcohols. While, *T, D, M, G, MeOH, MeOl* refers to Triolein, Diolein, Monolein, Glycerol, Methanol and Methyl Oleate respectively in Equations 1-3.

$$\mathrm{T + MeOH \Leftrightarrow D + MeOl} \quad (1)$$

$$\mathrm{D + MeOH \Leftrightarrow M + MeOl} \quad (2)$$

$$\mathrm{M + MeOH \Leftrightarrow G + MeOl} \quad (3)$$

$$r_1 = \frac{k_{MeOH}\left([MeOH] - (1/K_{eq})([MeOl][D]/[T])\right)}{1 + (K_A/K_{eq})([MeOl][D]/[T]) + K_A[D] + K_A[M] + K_A[G]} \quad (4)$$

$$r_2 = \frac{k_{MeOH}\left([MeOH] - (1/K_{eq})([MeOl][M]/[D])\right)}{1 + (K_A/K_{eq})([MeOl][M]/[D]) + K_A[D] + K_A[M] + K_A[G]} \quad (5)$$

$$r_3 = \frac{k_{MeOH}\left([MeOH] - (1/K_{eq})([MeOl][G]/[M])\right)}{1 + (K_A/K_{eq})([MeOl][G]/[M]) + K_A[D] + K_A[M] + K_A[G]} \quad (6)$$

However, equal rate constants for all three consecutive heterogeneous reactions (Equations 1-3) are significantly different from the kinetic constants and activation energies obtained for the homogeneous reaction systems. In view of this, we have developed a more detailed and comprehensive kinetic mechanism and model for heterogeneously catalyzed transesterification reactions.

Kinetic Mechanism

Cantrell et al.[8] developed a series of hydrotalcite catalyst with the general formulae of $[Mg_{(1-x)}Al_x(OH)_2]^{x+}(CO_3)_{x/n}^{2-}$ with x in the range 0.25–0.55. Transesterification reactions were performed in stirred batch reactor at 333 K using 0.01 mol (3 cm^3) of glyceryl tributyrate and 0.3036 mol (12.5 cm^3) methanol. The batch reactor was run for 3 hours.

To evaluate the industrial applicability of the heterogeneous catalyzed process, we have developed a new kinetic model based on the information obtained form Kinetic Monte Carlo (KMC) simulations. The elementary reactions were assumed according to 'Eley-Rideal' type kinetic mechanism (Table 1) elementary reaction. The first step is the adsorption of methanol. The adsorbed methanol then reacts with triolein, diolein or monolein to form diolein, monolein, and glycerol respectively along with methyl oleate. Equilibrium between adsorbed and desorbed species regulates the bulk and surface phase concentration for diolein, monolein and glycerol respectively. Each of these steps can be treated as a rate determining step.

Table 1: Biodiesel Reactions (Eley-Rideal Mechanism)

$* + CH_3OH \Leftrightarrow CH_3OH*$	**(a)**
$CH_3OH* + T \Leftrightarrow D* + MeOl$	**(b)**
$CH_3OH* + D \Leftrightarrow M* + MeOl$	**(c)**
$CH_3OH* + M \Leftrightarrow G* + MeOl$	**(d)**
$D* \Leftrightarrow D + *$	**(e)**
$M* \Leftrightarrow M + *$	**(f)**
$G* \Leftrightarrow G + *$	**(g)**
	7

Here, *, CH_3OH*, $D*$, $M*$, $G*$ represents empty site, methanol, diolein, monolein, and glycerol adsorbed respectively. *T*, CH_3OH, *D*, *M*, *G*, *MeOl* represents triolein, methanol, diolein, monolein, glycerol, and methyl oleate respectively. The overall reaction rate expressions have been derived based on the assumption of quasi steady state conditions for the surface species. We assumed that the concentrations of the catalyst surface species remained constant with respect to time and all the forward reactions rate constants were greater than the backward reaction rate constants. The above elementary reaction equations in Table 1 lead to the following kinetic rate expressions.

Table 2: Kinetic reaction rate expression

Rate of consumption of T	$R_T = \left(\dfrac{k(2)[T]}{1 + \left(\dfrac{k(2)[T] + k(3)[D] + k(4)[M]}{k(1)[CH_3OH]} \right) + \dfrac{k(2)[D]}{k(6)} + \dfrac{k(2)[T]}{k(5)} + \dfrac{k(4)[M]}{k(7)}} \right)$	**8**
Rate of consumption of D	$R_D = \left(\dfrac{k(3)[D] - k(2)[T]}{1 + \left(\dfrac{k(2)[T] + k(3)[D] + k(4)[M]}{k(1)[CH_3OH]} \right) + \dfrac{k(2)[D]}{k(6)} + \dfrac{k(2)[T]}{k(5)} + \dfrac{k(4)[M]}{k(7)}} \right)$	**9**
Rate of consumption of M	$R_M = \left(\dfrac{k(4)[M] - k(3)[D]}{1 + \left(\dfrac{k(2)[T] + k(3)[D] + k(4)[M]}{k(1)[CH_3OH]} \right) + \dfrac{k(2)[D]}{k(6)} + \dfrac{k(2)[T]}{k(5)} + \dfrac{k(4)[M]}{k(7)}} \right)$	**10**
Rate of consumption of Methanol	$R_{Meth} = \left(\dfrac{-(k(5)[D] + k(3)[T] + k(7)[M])}{1 + \left(\dfrac{k(3)[T] + k(5)[D] + k(7)[M]}{k(1)[CH_3OH]} \right) + \dfrac{k(5)[D]}{k(11)} + \dfrac{k(3)[T]}{k(9)} + \dfrac{k(7)[M]}{k(13)}} \right)$	**11**
Rate of change for Glycerol	$R_G = \left(\dfrac{k(7)k(1)[CH_3OH]}{k(3)[T] + k(5)[D] + k(7)[M]} \right)$	**12**

Rate of generation of Methyl Oleate	$$R_{MeOl} = \left(\frac{(k(5)[D] + k(3)[T] + k(7)[M])}{1 + \left(\frac{k(3)[T] + k(5)[D] + k(7)[M]}{k(1)[CH_3OH]} \right) + \frac{k(5)[D]}{k(11)} + \frac{k(3)[T]}{k(9)} + \frac{k(7)[M]}{k(13)}} \right)$$	**13**

Here, *k(i)* represents the kinetic rate for the forward reaction i in Table 1. $[D],[T],[M],[G],[CH_3OH],[MeOl]$ are the bulk phase activities for diolein, triolein, monolein, glycerol, methanol and methyl oleate respectively.

Determination of kinetic rate constants using genetic algorithm

The concentration (C_i) ($\forall i \in T, D, M, G, MeEs, MeOH$) profile in a stirred batch reactor is a function of kinetic rates (R_i) and is given by the following equation.

$$\frac{dC_i}{dt} = R_i \qquad \textbf{14}$$

This set of ordinary differential equations is solved by ordinary differential equations solver ode45 in MATLAB. The concentrations profile of components as a function of time predicted using MATLAB solver are the inputs to the genetic algorithm[10], which is used to determine the kinetic rate constants. The objective function is defined as the residual sum of squares of difference between the experimental and simulated values for concentration in a stirred batch reactor. It is calculated as the sum of the square of the difference between the predicted and the experimental values of the concentrations of the components (I = 1 to ncomp), at subsequent time points (j = 1 to *ntime*) at which the experimental data are available.

$$RSSQ = \sum_{j=1}^{ntime} \sum_{i=1}^{ncomp} \left(C_{i,j} - C_{i,j,epxerimental}\right)^2 \qquad \textbf{15}$$

The algorithm for the determination of the kinetic rate constants can be described as follows[11] (Figure 1)

- Specify the initial concentration to the batch reactor (C_{io}), size of the reactor, and batch time as the inputs to the genetic algorithm. The bounds of rate constants are also provided.
- Initial N chromosomes are generated, with each chromosome representing a feasible solution in terms of decision variables (rate constants – k(i)).
- The rate constants and initial concentration are input into the reactor model. The concentration profile with respect to time is generated as an output. The activity for the components $[i]$ ($\forall i \in T, D, M, G, MeEs, MeOH$) is a function of the concentration (C_i) and the activity coefficient Ψ_i.
- The chromosomes are ranked based on the objective function.

- The best x percent (this x is parameter for genetic algorithm) chromosomes are kept for the next iteration.
- Generate a new set of chromosomes by mutation, crossover in remaining chromosomes. The cost of the new chromosomes is evaluated on the objective function.
- Simulations are repeated until RSSQ (Residual sum of squares) is a very low value within the expected tolerance limits (RSSQ<0.005) or the maximum number of iterations (100000) is exceeded.

The GA algorithm used in this simulation was given by Haupt and Haupt[10]. The free software given by them as a part of the book is used for genetic algorithm.

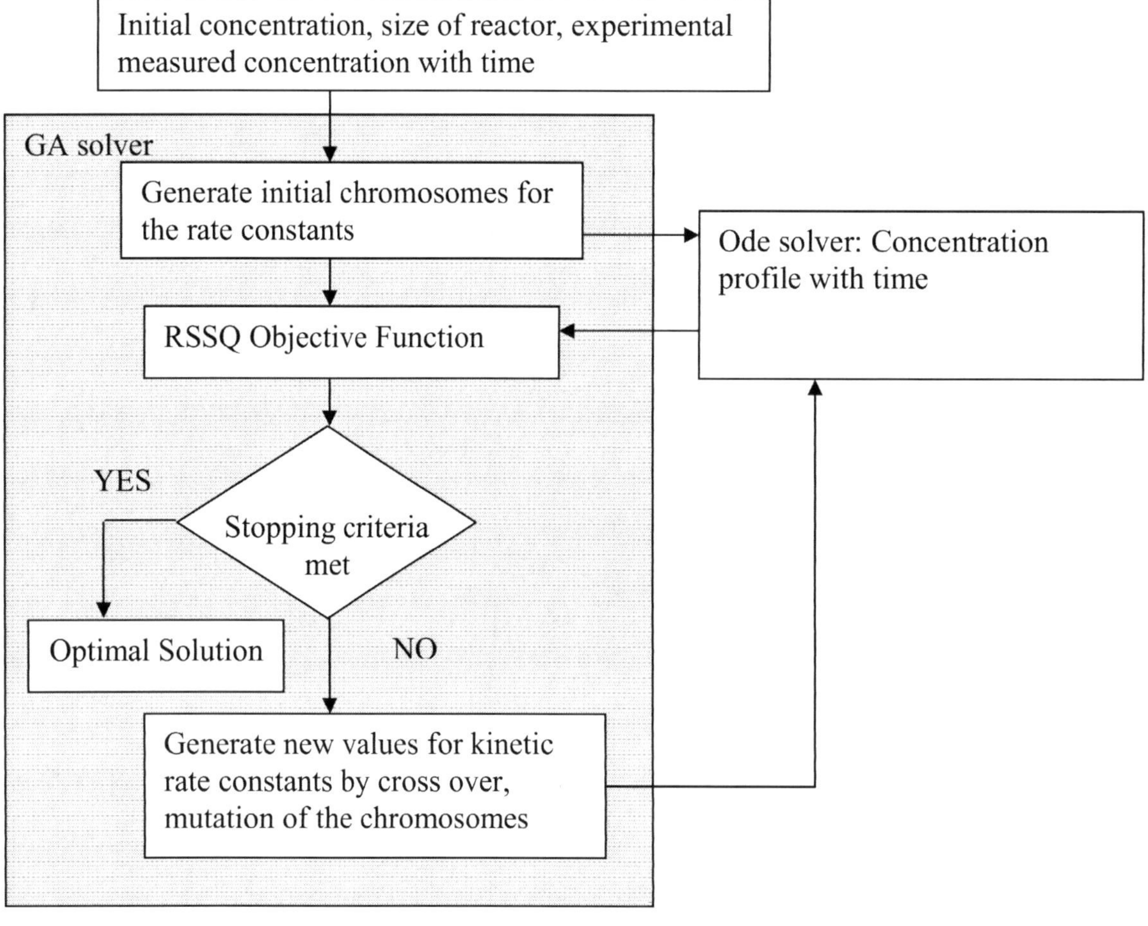

Figure 1: Algorithm for determination of kinetic parameters

Hybrid Monte Carlo/Mean field approach: By including the effect of surface reaction and surface species concentration in the bulk simulation we can make our simulation model more robust. The elementary reactions can be properly accounted for and the assumptions made in the derivation of the kinetic model can be justified. Hence in this work we presented a unique hybrid model combining Monte Carlo for the surface and mean field simulation for the bulk. The algorithm for this is given in Figure 2. The algorithm is again implemented in MATLAB.

1. Initial input to the Monte Carlo included the concentration of species on the surface of the catalyst, and the mole fraction in the bulk. The surface of the catalyst is considered to be empty at the start of a reactor run. The bulk concentrations were obtained from the work of Cantrell et al.[8].
2. For a small time interval dt(~1s) the surface concentration is assumed to be constant in the mean field simulation for the bulk phase. The bulk concentration changes were updated by the mean field simulation of the ordinary differential equation in the bulk phase.
3. Similarly bulk concentration is assumed to be constant over a small time interval dt for the MC simulation on the catalyst surface. The surface concentration was advanced in time by MC simulation.
4. The overall time is advanced as the times for MC and mean field simulation were updated.
5. The simulation is continued for the total run time of a reactor. (in this case 10800s).

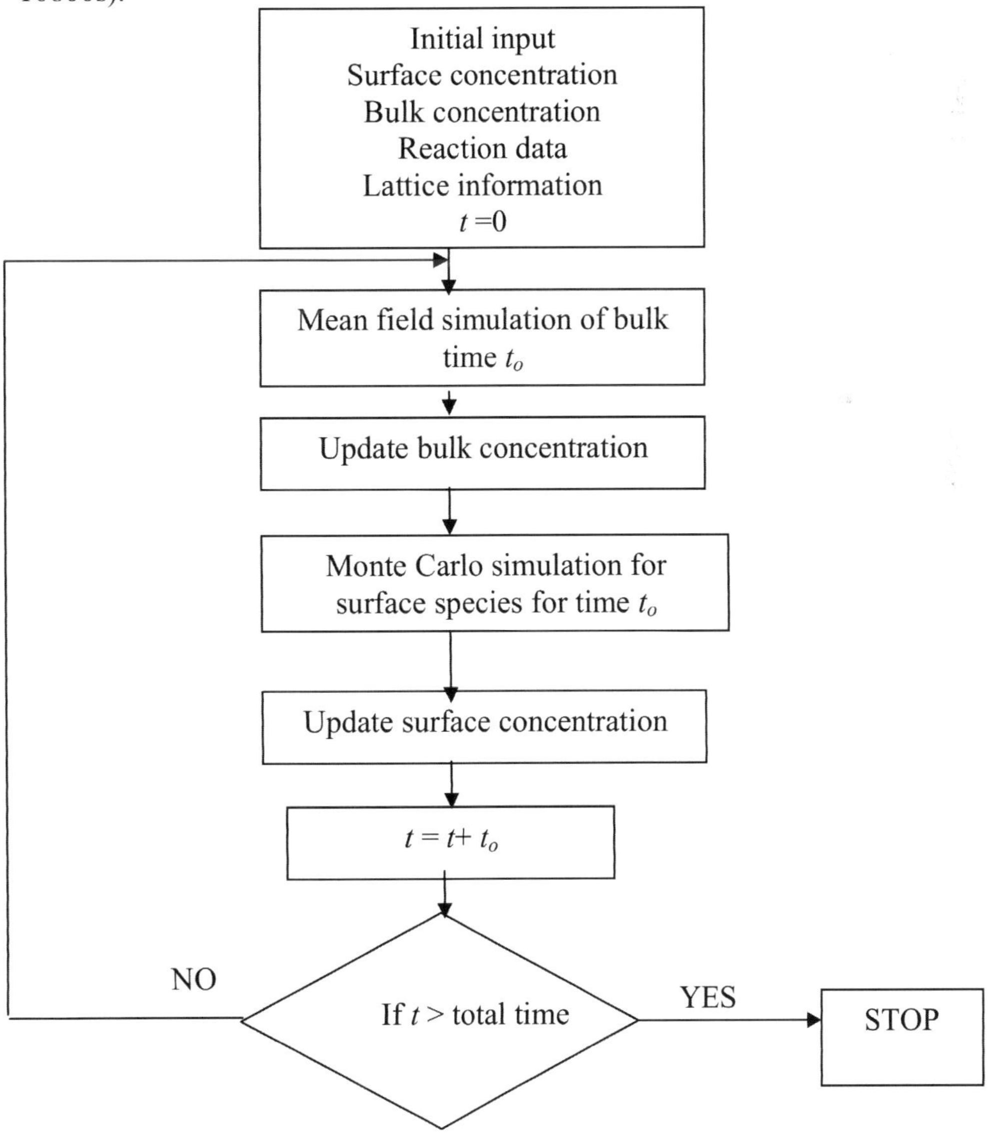

Figure 2: Hybrid Monte Carlo/mean field approach

The MC simulation implemented in MATLAB is based on the concept of class of reactions, categorised depending on the number of neighbouring reactants participating in a reaction[1]. The probability of each reaction at a lattice site is based on the occupancy of the neighbouring sites. This information about the reaction and their positions on lattice is stored in a variable called class. For a two dimensional lattice there are four possible classes. If a reaction is possible in one, two, three or four adjacent sites it can be called as class one, two, three or four respectively.

Kinetic Monte Carlo (MC) Algorithm[1]

1. The lattice is initialized at the start of a run. At time t=0 the catalyst surface is assumed to be empty. At other times the lattice is initialized with the distribution of species resulted from the previous simulation.
2. The whole lattice is searched for the possible reactions and classes. This information is stored in a variable called class which is a function of the class number, reaction and its position on the lattice.
3. Reaction is possible if the reactants are present on the adjacent site in the lattice. This information is already stored in the variable class. Hence the probability of a reaction is calculated based on the reaction rate constants obtained in previous algorithm, and its class information.
4. A reaction is selected based on its probability calculated in the above step using an inbuilt random number generator algorithm in MATLAB. Similarly for this reaction, we determine the possibility of selecting a site with one, two, three or four possible neighbouring reaction sites.
5. With this reaction and class information, a site/pair on the lattice surface is selected.
6. The class information and lattice surface distribution are updated locally.
7. The time is advanced with an average reaction time calculated in step 3.
8. The process is continued till the maximum run time t_0 is reached. In our case this t_0 is the same as the small update time dt used in the hybrid algorithm.

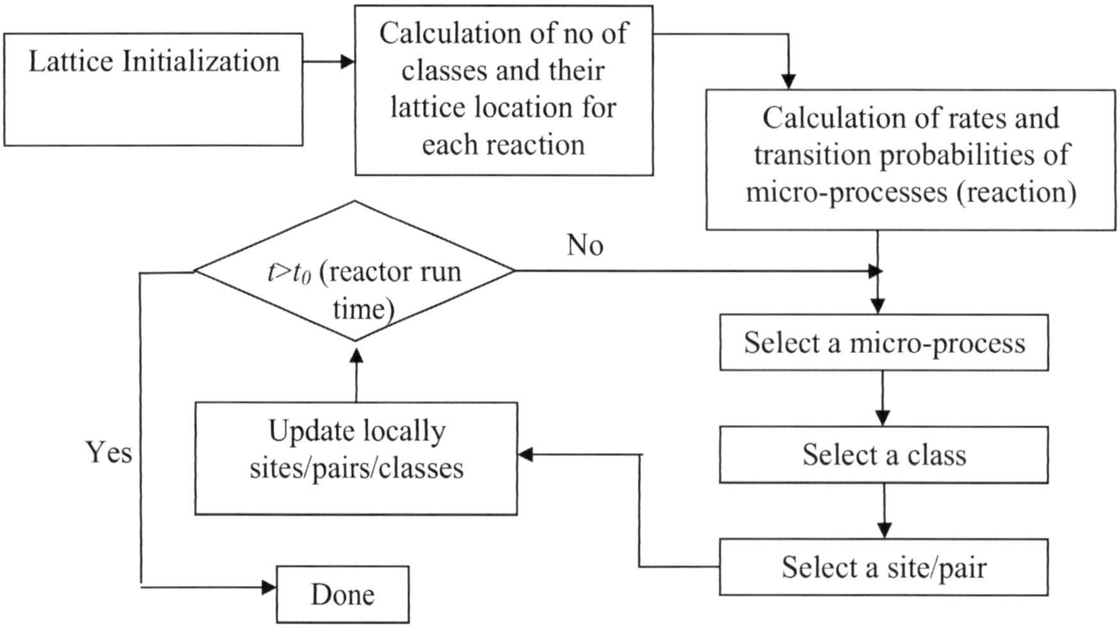

Figure 3: Continuous time MC algorithm[1]

Solution strategy

The following assumptions were made for the simulation of the above model. The reactor was considered to be isothermal. The molar volume and the total volume were assumed to be constant and there was no catalyst deactivation. The resulting system was simulated on a PC with Pentium® D CPU 3.00 GHz processor and 1GB RAM. The run time for the simulation was 105 minutes. The process conditions and parameters are taken from the paper of Cantrell et al.[8]

Results and Discussions

Figure 4 shows the comparison of the experimental data (points) with the simulated results (solid line). The kinetic model and the parameters described the biodiesel process well. The simulation results matched closely with the experimental results, with a low residual sum of squares. The kinetic parameters resulted are given in Table 3.

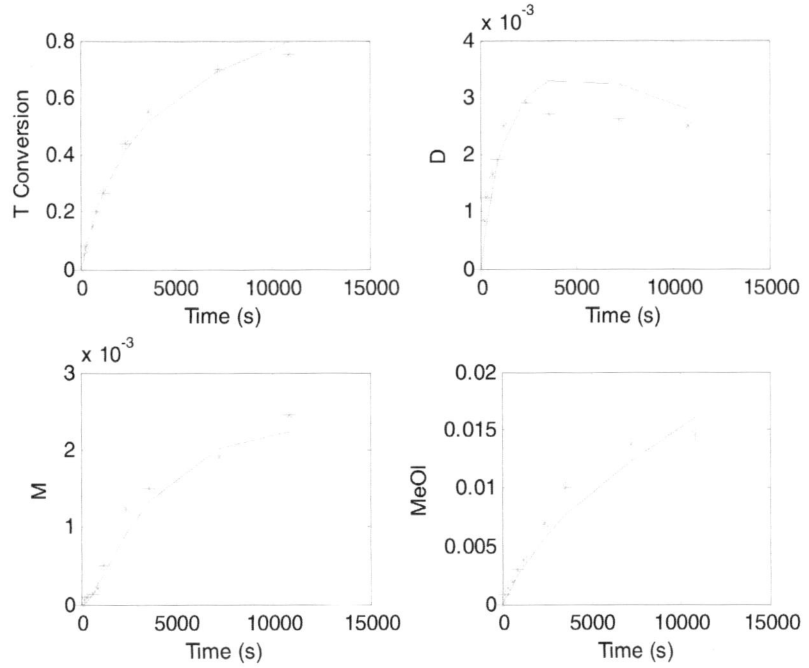

Figure 4: Comparison between experimental data and kinetic model for conversion of T and moles of D, M, and MeOl respectively

Table 3: Kinetic parameters for hyrotalcite

$k(1)$ $m^3 kg_{cat}^{-1} s^{-1}$	0.0001
$k(2)$ $m^3 kg_{cat}^{-1} s^{-1}$	0.039
$k(3)$ $m^3 kg_{cat}^{-1} s^{-1}$	0.1741
$k(4)$ $m^3 kg_{cat}^{-1} s^{-1}$	0.3587
$k(5)$ $m^3 kg_{cat}^{-1} s^{-1}$	0.0564

$k(6)$ $m^3 kg_{cat}^{-1} s^{-1}$	0.1579
$k(7)$ $m^3 kg_{cat}^{-1} s^{-1}$	0.0082
RSSQ	0.004385

The hybrid mechanism provided us with an understanding of the kinetic mechanism on the catalyst surface. The following assumptions were made to derive this mechanism

1. A quasi steady state for surface species. The concentration of components on the surface of the catalyst was found out to be almost constant. Only major change is in the empty surface concentration at the start of the reaction. After that the surface concentration remains constant with time.
2. Forward reactions are faster than the backward reactions. The reaction rates as determined by MC simulations show that the forward reaction is considerably faster than the backward reactions. Hence backward reactions can be neglected in the determination of final rate expression.

Conclusion

Hydrotalcite is a viable heterogeneous catalyst for biodiesel production. High conversion of Triolein is obtained and glycerol is formed as the main by product. Hence a more environmentally friendly chemical process for biodiesel production can be obtained since base removal and product separation are easier than in the traditional processes. A kinetic model based on Eley-Rideal type elementary reaction mechanism has been observed to give a satisfactorily description of the experimental data. The kinetic model is based on the assumption of quasi steady state for the surface species. Activity coefficient is used to account for the non-ideal behaviour in this analysis. The surface coverage of different species on the catalyst plays an important role in the reaction kinetics. . To account for the effect of the surface coverage on catalysts Monte Carlo simulations were performed. Further mean field simulation of the bulk phase was combined with the surface MC simulation in order to capture both the changes in the concentration profiles in the bulk as well as on the catalyst surface, simultaneously. Once the catalyst activity and composition and reactor configuration and geometry can be explicitly taken into account in the MC and mean field simulation models respectively, the proposed hybrid framework can be used for the design and decision making of the catalyst and the process.

Nomenclature

$[i]$	activity of each species i, mol/m^3
C_i	concentration of each species i, mol/m^3
K_a	adsorption equilibrium constant of the alcohols present, $m^3 mol^{-1}$
K_{eq}	equilibrium coefficient of the overall transesterification reaction, dimensionless
k_{MeOH}	rate coefficient for methanol adsorption, $m^3 kg_{cat}^{-1} s^{-1}$
$k(i)$	Reaction rate constant, $m^3 kg_{cat}^{-1} s^{-1}$ for forward reaction i in Table 1

ntime	total number of experimental values, dimensionless
ncomp	No of components
r_i	reaction rate for ith reaction, $molkg^{-1}s^{-1}$
R_i	reaction rate of the jth component, $molkg^{-1}s^{-1}$
t	time, s
γ_i	activity coefficient of species i, dimensionless
*	Empty surface site
*CH3OH**	Methanol adsorbed
*D**	Diolein adsorbed
*M**	Monolein adsorbed
*G**	Glcyerol adsorbed
T	Triolein bulk conc, mol/m^3
D	Diolein bulk, mol/m^3
M	Monolein bulk, mol/m^3
G	Glycerol bulk, mol/m^3
MeOl	Methyl Oleate bulk, mol/m^3
MeOH	Methanol

References

1. Reese JS, Raimondeau S, Vlachos DG. Monte Carlo Algorithms for Complex Surface Reaction Mechanisms: Efficiency and Accuracy. *Journal of Computational Physics.* 2001;173(1):302-321.
2. Narasimharao K, Lee A, Wilson K. Catalysts in production of biodiesel: A review. *Journal of Biobased Materials and Bioenergy.* Apr 2007;1(1):19-30.
3. Dossin TF, Reyniers MF, Marin GB. Kinetics of heterogeneously MgO-catalyzed transesterification. *Applied Catalysis B-Environmental.* 2006;62(1-2):35-45.
4. Dossin TF, Reyniers MF, Berger RJ, Marin GB. Simulation of heterogeneously MgO-catalyzed transesterification for fine-chemical and biodiesel industrial production. *Applied Catalysis B-Environmental.* 2006;67(1-2):136-148.
5. MacLeod CS, Harvey AP, Lee AF, Wilson K. Evaluation of the activity and stability of alkali-doped metal oxide catalysts for application to an intensified method of biodiesel production. *Chemical Engineering Journal.* 2008;135(1-2):63-70.
6. Abreu FR, Lima DG, Hamu EH, Wolf C, Suarez PAZ. Utilization of metal complexes as catalysts in the transesterification of Brazilian vegetable oils with different alcohols. *Journal of Molecular Catalysis A: Chemical.* 2004;209(1-2):29-33.
7. Tanabe K, Misono M, Ono Y, Hattori H. *Stud. Surf. Sci. Catal.* Vol 51.; 1989.
8. Cantrell DG, Gillie LJ, Lee AF, Wilson K. Structure-reactivity correlations in MgAl hydrotalcite catalysts for biodiesel synthesis. *Applied Catalysis a-General.* 2005;287(2):183-190.
9. Watkins RS, Lee AF, Wilson K. Li-CaO catalysed tri-glyceride transesterification for biodiesel applications. *Green Chemistry.* 2004;6(7):335-340.

10. Haupt R, Haupt SE. *Practical Genetic Algorithms* 2ed: John Wiley & Sons Inc 2004.
11. Mian X. *Biofuels production and optimization* [PhD Thesis], The University of Manchester; 2009.

85d Quantifying Enzyme Activity Using Quartz Crystal Microbalance with Dissipation Monitoring (QCM-D)

Mark A. Poggi and Matthew Dixon. Q-Sense, 808 Landmark Drive, Suite 124, Glen Burnie, MD 21061

There is a growing need for new technologies to quantitatively measure the surface properties of biological and material interfaces. One particular technique: Quartz Crystal Microbalance with Dissipation Monitoring (QCM-D) measures several dynamic properties of materials and interfaces. Using QCM-D, changes in mass and viscoelasticity can be simultaneously monitored in real-time in liquid. This technique has allowed scientists to achieve a more fundamental understanding regarding how dynamic biological, polymeric and other nano-scale materials behave on the molecular scale. This presentation will review recent advances regarding how we are currently using this technology to quantify the real-time activity of enzymes since enzymes are integral components in the conversion of biomass to liquids.

Web Page: www.q-sense.com

85e Solvents for Oil Extraction and Biodiesel Production

Arunprakash T. Karunanithi, Department of Civil Engineering, University of Colorado Denver, 1200 Larimer Street, Denver, CO 80217

Solvents are used to extract oil from the feedstock for biodiesel production. Appropriate selection of solvents is crucial for maximum oil extraction. This presentation introduces the application of solvent database search techniques and state of the art computer-aided molecular design techniques for solvent selection for oil extraction. A promising approach to convert the extracted oil into biodiesel is to use a co-solvent during transesterification reaction. The co-solvent increases the reaction rate at ambient operating conditions by forming a single phase solution of the reactants. Since the reaction can be carried out at ambient temperature and pressure the energy use for the biodiesel production process is greatly reduced. This presentation also discusses appropriate solvent selection strategies for this process.

98a Waste Biomass Gasification Characteristics in a Fixed Bed Gasifier

See Hoon Lee, Young Il Son, Chang Bok Go, Kyung Bin Choi, and Jae Ho Kim. Gasification research center, KIER, 71-2 Jang-dong Yusung-gu, Daejeon, South Korea

The gasification processes to convert low grade solid fuels into valuable fuels have been used in various applications, but the biomass gasification has only been used in small scale processes because the quantity of biomass supply generally is unstable. However the biomass gasification processes is widely developed due to the highly price of oil and the concern of climate change. Using results from the basic data and coal gasifier experiments, 100kg/hr(2ton/day) fixed bed gasifier was design and constructed. The pilot scale woody biomass gasification system consisted of feeding system, fixed bed gasifier with rotary stoker, oxidant feeding system, preheater, gravity fine particle collector, heat exchanger for syngas cooling, ID fan, and cooling tower. To monitor gasification reaction conditions, IR analyzer and data acquisition system was equipped. At the 5 times gasification experiments, the calorific value of syngas was 1,100~1,500 kcal/Nm3. At the gas engine experiments by using syngas, engine operation was stable when CO/H2 ratio was lower than 1. Because the calorific value might be relatively low, the injection volume of engine had to increase. Compared with LPG, the maximum combustion pressure was lower than 5bar. By using previous experimental results 20kw gas engine was equipped and tested.

Process-product studies on pyrolysis oil upgrading by hydrotreatment with Ru/C catalysts

A.R.Ardiyanti, R.H.Venderbosch, H.J.Heeres

1. Introduction

Biomass is considered an attractive alternative source for fossil derived energy carriers, liquid transportation fuels and bulk-chemicals. Liquid energy carriers can be obtained by biomass liquefaction followed by further upgrading. A promising liquefaction approach is fast pyrolysis. The product, pyrolysis oil, is a viscous liquid, with a high oxygen content (up to 40 wt%). The latter renders the oil polar and acidic due to the presence of organic acids. Various upgrading technologies have been proposed to improve the product properties to extend the application range. An attractive approach to extend the application range of pyrolysis oil is by its co-feeding into existing refineries, either in hydrotreating or FCC units. However, pyrolysis oil as such is not directly suitable for this purpose, because the oil is not miscible with hydrocarbon feedstocks and, moreover, shows a high tendency for coking, leading to blockage of feeding lines and reactors.

A promising upgrading technique is catalytic hydrotreating using hydrogen in combination with heterogeneous catalysts[1]. Typically, the objective is to obtain upgraded oils with very low residual oxygen contents to mimic the conventional crude oils. A major drawback of this approach is the high hydrogen consumption, which has a strong negative impact on the process economics[2].

An alternative is a mild hydrotreating process at a much lower severity than the conventional hydrotreating processes. Here we report experimental studies on the hydrotreating of pyrolysis oil at different severities in bench scale continuous packed bed reactors, addressing oxygen content, water content, average molecular weight and the tendency of coking of the products obtained.

2. Experiment Section

Materials

Pyrolysis oil was obtained by fast pyrolysis of forest residue and the composition is given in Table 1. Ru/C was used as the hydrotreating catalyst (5 wt% Ru, pelletized). The hydrogen used was of 99.9% purity.

Table 1 Elemental composition of the pyrolysis oil used in this study

Elements	Composition (wt%)
C	40.1
H	7.6
N	0.1
O (by difference)	52.1
O (dry basis)	40.1
Water	23.9

Reactor set-up

A schematic representation of the set-up used is given in Figure 1. The main part of the set-up is four tubular reactors-in-series filled with catalyst. The temperature of each reactor can be controlled independently.

3. Typical description of a hydrogenation experiment

Pyrolysis oil was pumped through the reactors. After stable liquid flow was obtained the hydrogen was fed into the reactor, then the pressure and temperature were raised to the desired conditions (200 bar of hydrogen pressure, other details see Table 1). The products (gas and liquid) were separated in a knock out vessel. The gas was vented and the liquid was collected in bottles. The temperature, pressure, flow of gas and liquid in and out of the reactors were recorded.

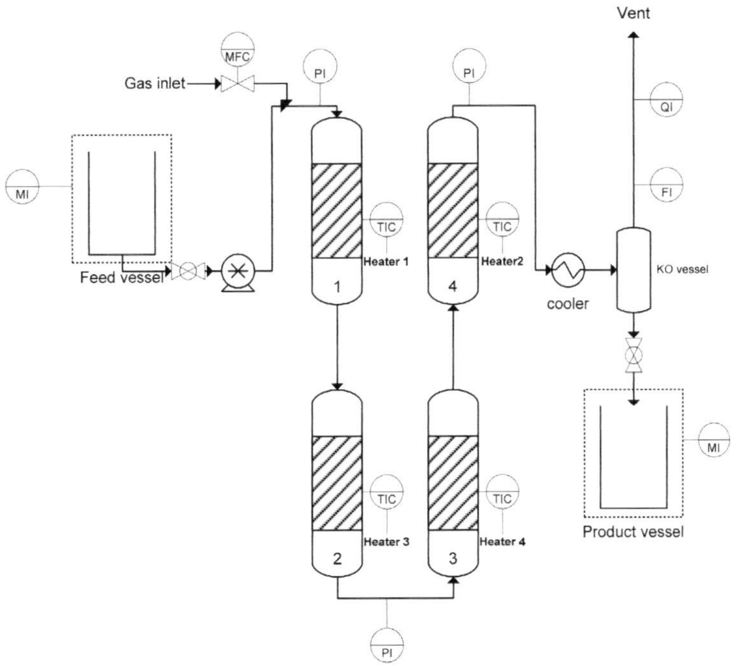

Figure 1 Experimental set-up used for pyrolysis oil hydrotreatment

3. Results and discussion

The experimental studies were carried out in a bench scale continuous set-up with four packed bed reactors in series (Figure 1). The temperature in each reactor is independently set at the pre-determined values. Ru/C was used as the catalysts and a fixed hydrogen pressure of 200 bar was applied. The pyrolysis oil feed rate and the temperatures were varied, see Table 1 for details.

Table 1 Experimental conditions

Code	Temperature (°C)[a]	WHSV (hr^{-1})	Process	Oxygen content (wt%, dry)
Py-oil				40.1
1	175-175-175-175	7.13	Stabilization	35.4
2	225-225-225-225	7.11	Stabilization	27.8
3	275-275-275-275	11.17	Stabilization	27.4
4	175-225-300-350	5.57	Mild hydrotreating	17.6
5	175-275-375-400	1.23	Mild hydrotreating	14.2
6	350-375-400-400	2.51	2-stage hydrotreating	12.3
7	350-375-400-400	2.35	2-stage hydrotreating	11.5

[a]temperature profile along 4 packed-bed reactors

For all experiments except run 1 two liquid products were obtained, viz. an aqueous phase and brown-red organic phase. The elemental composition, water content and average molecular weight of the organic product were determined. To gain insights in the coking tendency of the products, the samples were analyzed using TGA. The residual-weight at the end of the TGA scan was taken as a measure of coking, with a high residue indicating a high tendency for coking.

The oxygen content of the organic phase as a function of the process conditions is given in Figure 2, where the oxygen content of the product oil (dry basis) is depicted versus the severity of the experiments (higher severity means higher temperature and/or lower WHSV). Higher severity leads to lower oxygen contents. At low temperatures (175-275°C, stabilization step), the oxygen content reduces from 40.1% to 27.4 wt%.

A number of experiments were carried out at different temperatures for the 4 reactors (code 4-5). The first reactor in the set-up was set at 175°C, the last either at 350 or 400°C. At these conditions (mild hydrotreating), the oxygen content was reduced to about 14.2 wt%.

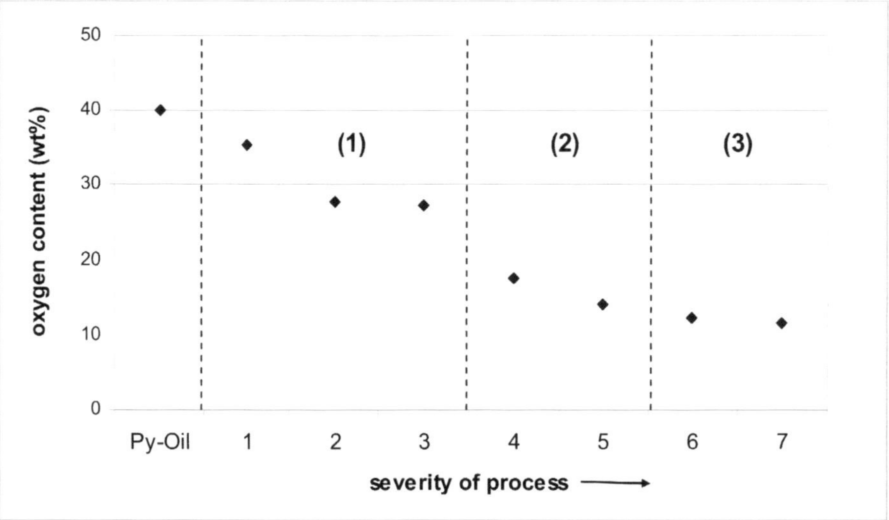

Figure 2 Oxygen content as function of process severity, (1) stabilization, (2) mild hydrotreating, and (3) 2-stage hydrotreating.

The product oil obtained at mild hydrotreating conditions (code 4, Table 1) was subjected to a second hydrotreating procedure at high temperatures (350-400°C, see code 6 and 7 in Table 1, 2-stage hydrotreating) as the oil was already stabilized. Also again two phases formed, viz. a black oil phase floating on top of a clear water layer. The oxygen content of the oil was reduced to 11.5 wt% at a WHSV of 2.35 h^{-1}.

The products of the stabilization steps (code 1-3, Table 1) were analyzed by GPC and TGA and the results are given in Figure 2. The molecular weight of the product increases upon going from 175°C to 275°C. Apparently, polymerization occurs during the stabilization step. This phenomenon has also been observed when heating up pyrolysis oil to 275°C in the absence of catalysts (HTTP process)[3]. The oxygen content of the product decreases, thus also hydrogenation occurs to some extent.

The residue after a TGA measurement is also a strong function of the temperature in the stabilization step. The residue is considerably higher at 275°C, and consequently has a higher tendency for coking and may be less suitable as a refinery feedstock. Of particular interest is the relation between the molecular weight of the products and the TGA residue. Products with a higher M_w also give higher TGA residues. This may be rationalized by assuming that the higher molecular weight fragments in the products are precursors for coke formation.

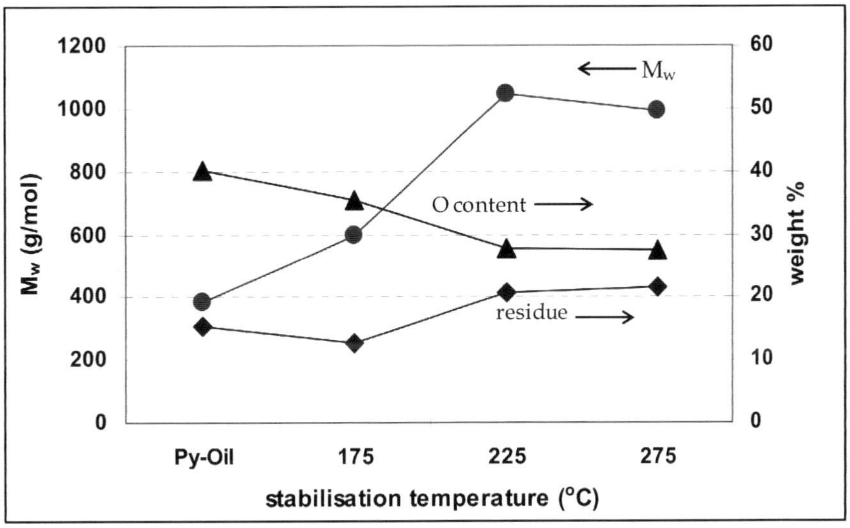

Figure 1 Oxygen content, mass average molecular weight and residual weight of organic phases obtained at 175-225°C (run 1-3 in Table 1).

The molecular weight and the TGA residue were also measured for the other samples in the series and the results are given in Figure 2. In this case, the oxygen content is taken as the independent variable, being an indicator for the severity of the process. For clarity, the results of the stabilization step are also provided.

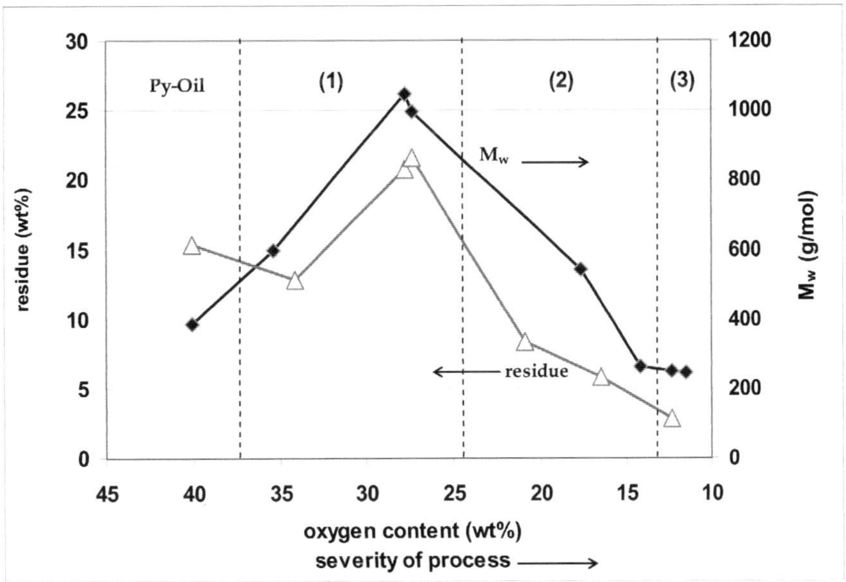

Figure 2 Mass average molecular weight and TGA residue of products from (1) stabilization, (2) mild hydrotreating, and (3) 2-stage hydrotreating.

The molecular weight of the products obtained at mild hydrotreating conditions (code 4-5, Table 1) are considerably lower than those produced in the stabilization phase (code 1-3). This suggests that the rate of molecular weight breakdown by hydrogenation or cracking reactions was on average much higher that the rate of polymerization. The TGA residue was also reduced considerably and again showed a clear relation with the molecular weight. The molecular break down continues during the 2-stage hydrotreating, resulting in organic products with even lower molecular weights and TGA residual weights than observed for product from the mild hydrotreatment procedure.

4. Conclusions and outlook

Hydrotreatment of pyrolysis oil at different severities in bench scale continuous packed bed reactors were performed. At temperature higher than 175°C the liquid product separated into an aqueous and organic phase. The oxygen content of the organic phase reduced proportionally with the severity.

During the stabilization experiments, polymerization (represented by increase of molecular weight) occurs faster than hydrogenation (represented by deoxygenation). Stabilization alone therefore appears not sufficient to make suitable product for refinery co-feeding.

During mild hydrotreatment (code 4-5, Table 1) and 2-stage hydrotreatment (code 6-7, Table 1) hydrogenation occurs faster than polymerization. This is clearly shown by a significant deoxygenation level and molecular weight break down, resulting in products with a lower tendency for coking than the products of the stabilization experiments. Clear relation between molecular weight and TGA residue of the products was observed. This implies that higher molecular weight fragments are precursors for coke formation. Actual FCC experiments with low-residual product (code 8, Table 1) as co-feed are in progress to verify whether the product is indeed suitable as a feedstock for refinery units.

References:

[1] Furimsky, E.; Appl. Catal. A: Gen., 2000, 12, 147-190

[2] Elliott, D.C.; Energy & Fuels 2007, 21, 1792-1815.

[3] Rep, M.; Venderbosch, R.H.; Assink, D.; Tromp, W.; Kersten, S.R.A.; Prins, W.; Van Swaaij, W.P.M.; 2006, Science in Thermal and Chemical Biomass Conversion; Bridgwater, A.V.; Boocock, D.G.B., Eds.; CPL Press: Chippenham, 1526-1535.

98c Influence of Coal-Biomass Feed Mixtures on Gasification Products

Nicholas C. Means[1], Paul Zandhuis[1], Götz Veser[2], Dirk Link[3], and Bryan D. Morreale[4]. (1) NETL Site Support Contractor, Parsons, 626 Cochrans Mill Road, Pittsburgh, PA 15129, (2) Department of Chemical Engineering, NETL Research Associate, University of Pittsburgh, Pittsburgh, PA 15261, (3) U. S. Department of Energy, National Energy Technology Laboratory, 626 Cochrans Mill Rd., P.O. Box 10940, Pittsburgh, PA 15236, (4) Office of Research and Development, U.S. Department of Energy, National Energy Technology Laboratory, 626 Cochrans Mill Rd., P.O. Box 10940, Pittsburgh, PA 15236

The United States Department of Energy's National Energy Technology Laboratory (NETL) is working to develop technologies that allow for the efficient conversion and utilization of domestically abundant resources in an effort to mitigate environmental issues, improve national security and develop greater fuel independence. Coal, an abundant natural resource, can be utilized by thermochemical processes to produce electricity, hydrogen, liquid fuels or other chemicals. Co-gasification of coal and non-food oriented biomass can prolong national fossil energy resources while reducing many environmental issues. Gasification can often be considered a two-stage process at low heating rates and a single stage process (simultaneous pyrolysis and gasification reactions) at high heating rates. Pyrolysis, a devolatilization reaction involving the breakdown of the macromolecular structure of carbonaceous fuel material and the release of volatile gases (CO, CO2, H2, CH4, H2O, etc.) and liquid products, first occurs during heat-up. After devolatilization is complete, the residual solid (char) is free to react by gasification reactions to produce syngas components. In order to further understand gasification phenomena, both reactions will be investigated at various feed compositions and reactor conditions. The focus of this paper is on the first and most complex pyrolysis reaction.

An experimental study on co-pyrolysis of Illinois#6 coal and switch grass was done in a batch reactor at conditions consistent with an entrained gasifier. Co-pyrolysis experiments were performed in an effort to gain an understanding of the effect of coal-biomass co-fed products on reaction kinetics and gaseous, liquid and solid product distributions. Coal and biomass were fed to the reactor with varying feed ratios (100%, 85%, 70%, 50% and 0% coal, balance biomass). Primary gaseous products (CO, CO2, CH4, H2 and H2O) were monitored and analyzed online using quadrupole mass spectrometry. Trace gaseous products were collected as a batch and analyzed using gas chromatography/mass spectrometry. Tar/liquid product analysis was done using gas chromatography/mass spectrometry and residual solid analysis was done with inductively coupled plasma optical emission spectroscopy. Initial results indicate that the addition of small quantities of biomass to coal can improve gaseous product yield and decrease product sulfur species.

Conversion of Pyrolysis Oil

Teerawit A. Prasomsri, Roberto E. Galiasso Tailleur, Lance L. Lobban, Richard G. Mallinson, and Daniel E. Resasco,

*Center for Biomass Refining and, School of Chemical, Biological and Materials Engineering, University of Oklahoma,
Norman, OK, USA*

Abstract

1- and 2- tetralones are used as representative compounds in biomass-derived pyrolysis oil (or bio-oil). This work focused on conversion of oxygenated aromatic compounds over HY zeolite. The reactions were conducted in a fixed-bed reactor at 300, 400 and 500 °C under hydrogen atmospheric pressure. Pure feeds were introduced to the reactor by using a bubbler in order to minimize condensation or solvent problems. It was found that 2-tetralone is thermally unstable and has a higher reactivity than 1-tetralone. Gaseous products were analyzed by on-line GC. Hydrogen abstraction (i.e. naphthol product) and coupling (i.e. heavies) were the main reaction paths. Deoxygenation (i.e. to tetralin and naphthol) products was a slower path.

Introduction

Transportation fuels predominantly come from conventional nonrenewable petroleum. Increasing uses of fuels from imported sources puts an emphasis in search for sustainable sources of energy. Biomass is a rational option for petroleum replacement since it is renewable, inexpensive, environmental friendly, and carbon rich. Biomass can be utilized in various manners depending upon the different types of feedstocks [Mohan *et at.*, 2006], such as direct/partial combustion, fixed/fluidized bed gasification, liquefaction, and fast pyrolysis [Briens *et al.*, 2008]. Biomass-derived pyrolysis oils (or bio-oils) are a complex mixture containing a high level of oxygen content. They are composed of various types of oxygenated compounds such as acids, aldehydes, furans, sugars, alcohol, ketones, as well as aromatics. Many works (reviewed by Huber and Corma, 2007) have studied catalytic cracking of various biomass-derived feedstocks under FCC conditions. Most of these studies concentrate on alcohols, ketones, acids, and aldehydes conversion. Oxygen removal and ring-opening are requisite in order to make the pyrolysis oil components more fungible with regular gasoline and diesel, as well as upgrading their properties. Hence, this work focused on the conversion of tetralones in order to study the aromatic effect on oxygenated compounds.

Experimental

Feed rate calibration

The molecules used in this study (tetralin, 1-tetralone, 1-tetralol, 1-naphthol, and 2-tetralone) are typically heavy and varied in boiling points. To avoid solvent usage and minimize condensation in the system, the feeds were introduced to the reactor by using a glass bubbler saturator (see Figure 1). Vapor concentration of individual compounds was modified by adjusting the heating bath temperature.

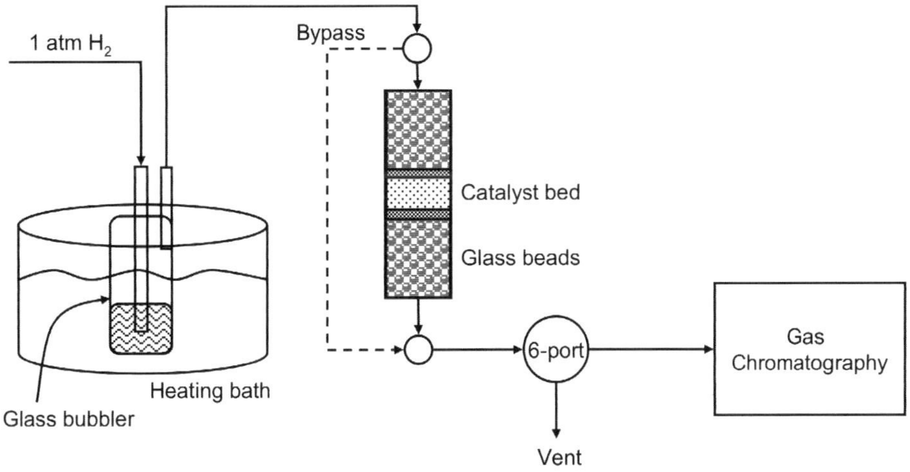

Figure 1 Schematic diagram of the system

Catalytic activity testing

The reactions were conducted in a ¼" stainless steel reactor. a commercial HY zeolite from Zeolyst International (CBV 780, Si/Al ratio 40) was used as the catalyst and packed in the center of the reactor. The top and bottom parts of the bed were filled with 3 mm-diameter glass beads. Operating conditions were atmospheric pressure in hydrogen, temperature range 300-500 °C, with various space velocities by changing the amount of catalyst. Prior to each run, the feed was directed through the bypass until the concentration reached steady-state, and then switched to the reactor. Products were analyzed periodically on-line with an HP5890 gas chromatograph equipped with HP-5 column and an FID detector. The temperature program started at 40 °C, linearly increased to 280 °C with a 15 °C/min heating ramp, and finally held for 8 min.

Results and Discussion

Thermal Decomposition

Thermal reactions of 1- and 2-tetralone were tested in a blank reactor (i.e. filled with glass beads and no catalyst) over the 300-500 °C temperature range. Conversion increased, as temperature increased for both compounds (Table 1). Selectivity towards hydrogen abstraction is higher than that towards deoxygenation (i.e. tetralones were mainly converted to naphthol rather than naphthalene). No ring-contraction (RC) or ring-opening (RO) products were observed in any of these cases. Interestingly, 2-tetralone exhibited significantly higher conversion at the same reaction temperature, which suggests that 1-tetralone is more stable (i.e. zero conversion at 300 °C) due to its conjugated double bonds between carbonyl group and the aromatic ring, which stabilize the molecule structure.

Table 1 Thermal decomposition of 1- and 2 tetralones

Feed	1-Tetralone			2-Tetralone		
Temperature	300 °C	400 °C	500 °C	300 °C	400 °C	500 °C
Conversion	0.0%	13.2%	49.1%	18.9%	76.0%	98.6%
Product Yields (wt%)						
RO	0.0%	0.0%	0.0%	0.0%	0.0%	0.0%
RC	0.0%	0.0%	0.0%	0.0%	0.0%	0.0%
Tetralin	0.0%	0.0%	3.1%	1.0%	1.9%	0.0%
Dihydronaphthalene	0.0%	0.0%	0.0%	0.0%	0.0%	0.0%
Naphthalene	0.0%	0.0%	3.5%	0.5%	2.1%	13.8%
Tetralone	100.0%	86.8%	50.9%	81.1%	24.0%	1.4%
Tetralol	0.0%	0.0%	0.0%	0.0%	0.0%	0.0%
Naphthol	0.0%	13.2%	41.7%	9.0%	69.0%	79.6%
Heavies	0.0%	0.0%	0.7%	8.2%	3.1%	5.2%

Reaction conditions: 1 atm H_2, TOS 1 h

Conversion of 1- and 2-Tetralone over HY Zeolite

The catalytic reaction was carried out at 300 °C over the HY zeolite under atmospheric pressure. Similar to the thermal reaction case, the conversion of 2-tetralone over HY zeolite (see Table 2) is higher than that of 1-tetralone.

Table 2 Product distribution of 1- and 2-tetralone conversions over HY zeolite

Feed	1-Tetralone	2-Tetralone
Conversion	77.1%	91.5%
Product Yields (wt%)		
RO	0.0%	0.0%
RC	0.0%	0.1%
Tetralin	4.4%	7.3%
Dihydronaphthalene	1.0%	0.0%
Naphthalene	11.8%	9.9%
Tetralone	22.9%	8.5%
Tetralol	0.0%	0.0%
Naphthol	33.2%	41.0%
Heavies	26.7%	33.1%
Product Selectivity (wt%)		
RO	0.0%	0.0%
RC	0.0%	0.2%
Tetralin	5.7%	8.0%
Dihydronaphthalene	1.3%	0.0%
Naphthalene	15.3%	10.8%
Tetralone	0.0%	0.0%
Tetralol	0.0%	0.0%
Naphthol	43.1%	44.8%
Heavies	34.6%	36.2%

Reaction conditions: 300 °C, 1 atm H_2, W/F 1.723 h, TOS 1 h

Also in this case, naphthol is the dominant product. In contrast to the blank test, the acid HY catalyst obviously enhances selectivity for deoxygenation of 1- and 2- tetralones, as well as formation of heavies [Noji *et al.*, 2003] which may arise from etherification of benzylic alcohols [Tarlani *et al.*, 2006 and Miller *et al.*, 2003]. However, as seen in the GC chromatogram, several peaks are lumped as heavies,. Product selectivity for 1- and 2-tetralone conversion is similar and is highest for hydrogen abstraction (i.e. naphthol), followed by coupling (i.e. heavies), and oxygen removal (i.e. tetralin and naphthalene) products.

Reaction Pathway of 1-Tetralone over HY Zeolite

To study the reaction pathway, experiments were conducted at 300 °C over HY zeolite under atmospheric pressure. Table 3 shows the product distribution (e.g., tetralin, 1-tetralol, and 1-naphthol) involved in the 1-tetralone reaction pathway (summarized in Figure 2).

Table 3 Product distribution of various feed reactions over HY zeolite

Feed	1-Tetralone	Tetralin	1-Tetralol	1-Naphthol
Conversion	77.1%	16.1%	100.0%	10.9%
Product Yields (wt%)				
RO	0.0%	0.0%	0.0%	0.0%
RC	0.0%	0.0%	0.0%	0.0%
Tetralin	4.4%	83.9%	42.6%	0.4%
Dihydronaphthalene	1.0%	0.0%	0.0%	0.0%
Naphthalene	11.8%	16.1%	57.4%	1.1%
Tetralone	22.9%	0.0%	0.0%	1.4%
Tetralol	0.0%	0.0%	0.0%	0.0%
Naphthol	33.2%	0.0%	0.0%	89.1%
Heavies	26.7%	0.0%	0.0%	8.0%

Reaction conditions: 300 °C, 1 atm H_2, W/F 1.723 h, TOS 1 h

In the presence of the acid catalyst HY, it is expected that the carbonyl group of 1-tetralone is protonated to form the oxonium ion, which is in resonance with the benzylic cation. Subsequently, a hydride migrates to the benzylic cation to give the benzyl alcohol [Vuligonda *et al.*, 1996], which is thermally unstable and easily dehydrated to form dihydronaphthalene. Naphthol possibly derives from tetralone as observed by Rebelo *et al.* [2005] and Yuranov *et al.* [2002] on metal catalysis, as well as the benzyl alcohol via hydride transfer step [Martan *et al.*, 1970]. As we have discussed above, coupling of naphthols results in heavies products.

Once, we used 1-tetralol as a feed, it is completely thermally decomposed to dihydronaphthalene and naphthalene with selectivity ~60% and ~35%, respectively. Unlike the blank test, we observed only tetralin and naphthalene with selectivity ~43% and ~57%, respectively, over HY. We suggest that 1-tetralol is thermally dehydrated to form dihydronaphthalene in the gas phase before it accesses acid sites, followed by disproportionation of the dihydronaphthalenes on the acid sites (i.e. intermolecular hydride transfer) to give naphthalene and tetralin [Corma *et al.*, 2001 and Santikunaporn *et al.*, 2004]. In addition, naphthalene also derives from the catalytic reaction of tetralin [Townsend and Abbot, 1992 and 1993]. As the temperature is increased, it may be expected that thermal equilibration among the hydrocarbon species will increase.

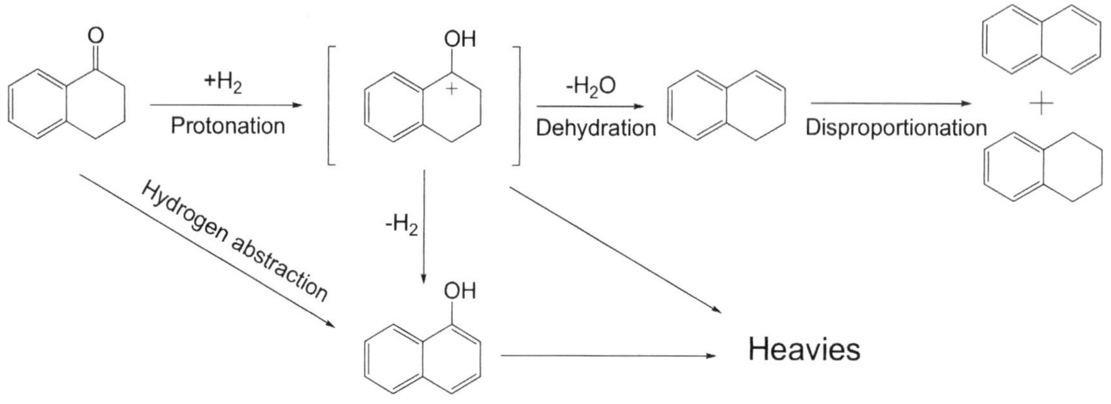

Figure 2 Proposed reaction pathway of 1-tetralone over HY Zeolite

Effect of temperature on the 1- and 2-tetralones conversions over HY

This series of experiments were studied in the temperature range 300 - 500 °C over the HY zeolite under hydrogen at atmospheric pressure (Table 4).

Table 4 Product distribution of 1- and 2-tetralone conversions with different temperatures

Feed	**1-Tetralone**		**2-Tetralone**		
Temperature	300 °C	400 °C	300 °C	400 °C	500 °C
Conversion	77.1%	81.8%	91.5%	99.5%	100.0%
Product Yields (wt%)					
RO	0.0%	0.0%	0.0%	0.0%	0.0%
RC	0.0%	0.9%	0.1%	1.9%	2.0%
Tetralin	4.4%	5.3%	7.3%	6.7%	2.6%
Dihydronaphthalene	1.0%	0.0%	0.0%	0.0%	0.0%
Naphthalene	11.8%	21.6%	9.9%	21.5%	22.9%
Tetralone	22.9%	18.2%	8.5%	0.5%	0.0%
Tetralol	0.0%	0.0%	0.0%	0.0%	0.0%
1-Naphthol	33.2%	42.2%	0.0%	0.0%	22.1%
2-Naphthol	0.0%	0.0%	41.0%	39.6%	40.8%
Heavies	26.7%	11.8%	33.1%	29.8%	9.7%
Product Selectivity (wt%)					
RO	0.0%	0.0%	0.0%	0.0%	0.0%
RC	0.0%	1.0%	0.2%	1.9%	2.0%
Tetralin	5.7%	6.5%	8.0%	6.7%	2.6%
Dihydronaphthalene	1.3%	0.0%	0.0%	0.0%	0.0%
Naphthalene	15.3%	26.4%	10.8%	21.5%	22.9%
Tetralone	0.0%	0.0%	0.0%	0.0%	0.0%
Tetralol	0.0%	0.0%	0.0%	0.0%	0.0%
1-Naphthol	43.1%	51.6%	0.0%	0.0%	22.1%
2-Naphthol	0.0%	0.0%	44.8%	39.8%	40.8%
Heavies	34.6%	14.4%	36.2%	30.0%	9.7%

Reaction conditions: 1 atm H_2, W/F 1.723 h, TOS 1 h

Conversions of both 1- and 2-tetralone increase at higher temperature. As the reaction temperature increases from 300 to 400 °C, selectivity to tetralin remains about the same (i.e. ~5% for 1-tetralone, and ~7% for 2-tetralone), whereas selectivity to naphthalene increases twofold (i.e. from ~15 to 30% for 1-tetralone, and from ~10 to 20% for 2-tetralone) since, over HY zeolite, tetralin can be converted to naphthalene [Townsend *et al.*, 1992]. Selectivity to naphthol and heavies are relatively the same at 300 °C for both 1-tetralone and 2-tetralone.

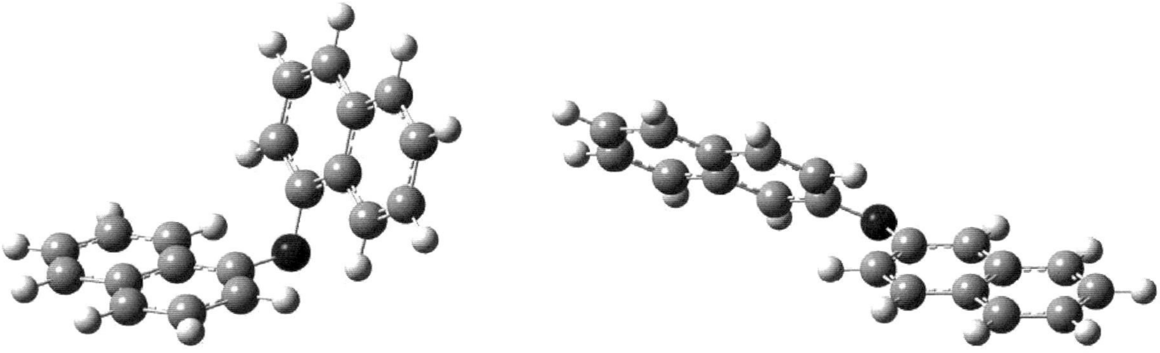

(a) (b)

Figure 3 Proposed structures of heavies with energy optimization are simulated by using GaussView. (a) from 1-tetralone reaction, (b) from 2-tetralone reaction

For 1-tetralone at 400 °C, selectivity to 1-naphthol increases since it tends to desorb quickly from the surface, which reduces the chance of forming heavy products. In contrast, we believe that 2-tetralones were produced at higher levels, but they were also consumed by coupling due to the structural constraint in which the flat adsorption (see Figure 3b) is energetically preferred [Yang *et al.*, 2003]. Therefore, when compared to 1-tetralone, we found higher selectivity towards heavies for 2-tetralone conversion at 400 °C. Interestingly, we are able to see isomerization of 1- and 2-naphthols via a carbocation intermediate of dihydronaphthalene oxide [Kiss *et al.*, 1999] or naphthalene oxide [Brandt *et al.*, 2002] at 500 °C. Furthermore, trace amounts of ring-contraction products were found for the acid catalyzed reaction [Santikunaporn *et al.*, 2007] at 400 and 500 °C. However, no ring-opening products were observed under the conditions investigated in this work.

Conclusion

1-tetralone is more stable than 2-tetralone due to the double bond of the carbonyl group being conjugated with the aromatic rings. At 300 °C, 2-tetralone can be thermally decomposed to mainly naphthol. However, selectivity to heavies, possibly resulting from coupling of naphthols, increases significantly in the presence of the acid catalyst HY. At higher temperatures and in the presence of acid sites, the 1- and 2-tetralone conversions are enhanced and give highest selectivity to hydrogen abstraction products, followed by condensation of heavies and deoxygenation products, respectively. Changes in selectivity are observed at higher temperature, 500 °C, together with isomerization of naphthols.

Acknowledgements

The authors are grateful for financial support provided by the State of Oklahoma OCAST, from the Secretary of Energy of the State of Oklahoma and the Oklahoma Bioenergy Center.

References

Brandt, P., Jia, Z. S., and Thibblin, A., (2002), Kinetic and thermodynamic stability of naphthalene oxide and related compounds: A comparative microcalorimetric and computational (DFT) study, *Journal of Organic Chemistry*, 67(22), 7676-7682.

Briens, C., Piskorz, J., and Berruti, F., (2008), Biomass valorization for fuel and chemicals production – a review, *International Journal of Chemical Reactor Engineering*, 6, Review R2.

Corma, A., Gonzalez-Alfaro, V., and Orchilles, D. V., (2001), Decalin and tetralin as probe molecules for cracking and hydrotreating the light cycle oil, *Journal of Catalysis,* 200, 34-44.

Huber, G. W. and Corma, A., (2007), Synergies between bio- and oil refineries for the production of fuels from biomass, Angewandte Chemie (International ed. in English), 46(38), 7184-7201.

Kiss, A., Jekõ, J., Lévai, A., and Patonay, T., (1999), Stereoselective α-oxyfunctionalisation of benzo(hetera)cyclanones by dimethyldioxirane, *Third International Electronic Conference on Synthetic Organic Chemistry (ECSOC-3)*, September 1-30

Martan, M., Manassen, J., and Vofsi, D., (1970), Oxidation of tetralin, α tetralol and α tetralone : Dependence of alcohol to ketone ratio on conversion, *Tetrahedron*, 26(15), 3815-3827.

Miller, K. J., and Abu-Omar, M. M., (2003), Palladium-catalyzed reactions of secondary benzylic alcohols: etherification, amination, and thioetherification, *European Journal of Organic Chemistry*, Issue 7, 1294-1299.

Mohan, D., Pittman, C., U., and Steele, P. H., (2006), Pyrolysis of wood/biomass for bio-oil: a critical review, *Energy Fuels*, 20(3), 848-889.

Noji, M., Ohno, T., Fuji, K., Futaba, N., Tajima, H., and Ishii, K., (2003), Secondary benzylation using benzyl alcohols catalyzed by lanthanoid, Scandium, and hafnium triflate, Jourmal of Organic Chemistry, 68(24), 9340-9347.

Rebelo, S. L. H., Simões, M. M. Q., Neves, M. G. P. M .S., Silva, A. M. S., Tagliatesta, P., and Cavaleiro, J. A. S., (2005), Oxidation of bicyclic arenes with hydrogen peroxide catalysed by Mn(III) porphyrins, *Journal of Molecular Catalysis A: Chemical*, 232 (1-2), 135-142.

Santikunaporn, M., Herrera, J. E., Jongpatiwut, S., Resasco, D. E., Alvarez, W. E., and Sughrue, E. L., (2004), Ring opening of decalin and tetralin on HY and Pt/HY zeolite catalysts, *Journal of Catalysis*, 228(1), 100-113.

Santikunaporn, M., Alvarez, W. E., and Resasco, D. E., (2007), Ring contraction and selective ring opening of naphthenic molecules for octane number improvement, Applied *Catalysis A: General*, 325(1), 175-187.

Tarlani, A., Riahi, A., Abedini, M., Amini, M. M., and Muzart, J., (2006), Catalytic condensation process for the preparation of organic peroxides from tert-butyl hydroperoxide and benzylic alcohols, *Applied Catalysis A: General*, 315, 150–152.

Townsend, A. T., and Abbot, J., (1992), Catalytic cracking of tetralin on HY zeolite, *Applied Catalysis A: General*, 90, 97-115.

Townsend, A. T., and Abbot, J., (1993), Catalytic cracking of tetralin on HZSM-5 zeolite, *Applied Catalysis A: General*, 95, 221-236.

Vuligonda, V., Lin, Y., and Chandraratna, R. A. S., (1996), Selective conversion of α-tetralones to dihydronaphthalenes, *Tetrahedron Letters*, 37(12),1941-1944.

Yang, H., Fairbridge, C., and Ring, Z., (2003), Adsorption of dibenzothiophene derivatives over a MoS_2 nanoclusterA Density Functional Theory study of structure−reactivity relations, Energy Fuels, 17(2), 387-398.

Yuranov, I., Kiwi-Minsker, L., and Renken, A., (2002), One-step vapour-phase synthesis of 2-methyl-1-naphthol from 1-tetralone, *Applied Catalysis A: General*, 226(1-2), 193-198.

AUTHOR INDEX

AUTHOR INDEX